THE MASTERING ENGINEER'S HANDBOOK

BOBBY OWSINSKI

FIFTH EDITION

The Mastering Engineer's Handbook
by Bobby Owsinski

Published by:
Bobby Owsinski Media Group
4109 West Burbank, Blvd.
Burbank, CA 91505

© Bobby Owsinski 2024
ISBN: 978-1-946837-42-4

ALL RIGHTS RESERVED. No part of this work covered by the copyright herein may be reproduced, transmitted, stored, or used in any form and by any means graphic, electronic or mechanical, including but not limited to photocopying, scanning, digitizing, taping, Web distribution, information networks or information storage and retrieval systems, except as permitted in Sections 107 or 108 of the 1976 Copyright Act, without the prior written permission of the publisher.

For permission to use text or information from this product, submit requests to office@bobbyowsinski.com.

Please note that much of this publication is based on personal experience and anecdotal evidence. Although the author and publisher have made every reasonable attempt to achieve complete accuracy of the content in this Guide, they assume no responsibility for errors or omissions. Also, you should use this information as you see fit, and at your own risk. Your particular situation may not be exactly suited to the examples illustrated herein; in fact, it's likely that they won't be the same, and you should adjust your use of the information and recommendations accordingly.

Any trademarks, service marks, product names or named features are assumed to be the property of their respective owners, and are used only for reference. There is no implied endorsement if we use one of these terms.

Finally, nothing in this book is intended to replace common sense, legal, medical or other professional advice, and is meant to inform and entertain the reader.

To buy books in quantity for corporate use or incentives, email office@bobbyowsinski.com.

Thanks So Much For Purchasing This Book!

Here's An Extra Free Bonus

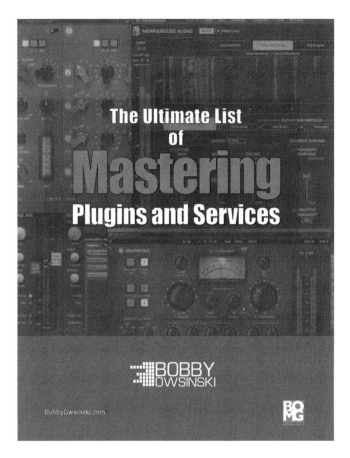

PDF Download
Download It Here

(use the QR Code or go to bobbyowsinskicourses.com/masteringreg)

TABLE OF CONTENTS

Introduction 1
    Meet the Mastering Engineers.. 2

PART I: THE MECHANICS OF MASTERING

The Essence of Mastering 7
    Why Master Anyway? 8
    From Vinyl, to CDs, to MP3s, and to Streaming... 9
    The Difference between You and a Pro.. 12
    Takeaways.. 14

Digital Audio Basics 15
    Sample Rate 15
    Bit Depth 17
    Standard Audio File Formats 18
    Data Compression 19
    Takeaways.. 20

Prepping for Mastering 21
    Mixing for Mastering.. 21
    Takeaways.. 24

Monitoring For Mastering 25
    The Acoustic Environment 25
    Let's Fix Your Listening Area 26
    The Monitors 27
    Listening Techniques For Mastering 34
    Takeaways.. 37

Mastering Tools 39
    The Compressor 39
    The Limiter 44
    The Equalizer 45
    The De-Esser.. 47

    Saturation 49

    Convertors 50

    Consoles/Monitor Control 51

    The Digital Audio Workstation 52

    Other Tools 53

    Takeaways 54

Metering Tools 57

    The Peak Meter 57

    The RMS Meter 59

    The Phase Scope 61

    The Phase Correlation Meter 64

    The Spectrum Analyzer 65

    The Dynamic Range Meter 67

    Dynamic Ranges of Different Genres of Music 68

    The LUFS Meter 69

    Takeaways 73

Mastering Techniques 77

    The Basic Mastering Technique 77

    Creating A Loud Master 78

    Frequency Balance Techniques 88

    The Mastering Signal Path 89

    Adding Effects 92

    Editing Techniques For Mastering 93

    Parts Production 94

    Multiple Masters 95

    Mastering Music For Film And Television 96

    Takeaways 97

Dedicated Mastering Plugins 99

    Dedicated Mastering Plugins 99

    AI-Assisted Mastering Plugins 103

    Where To Insert A Mastering Plugin 107

    Takeaways 108

Online Mastering Services 109
Pro Studio e-Mastering 109
Automated Online Mastering 110
Takeaways 114

Mastering For Online Distribution 115
Data Compression Explained 115
Creating Files For Streaming Services 121
Submitting To Online Streaming Services 124
Creating A FLAC File 126
Apple Digital Masters 126
Other High-Resolution Platforms 130
Takeaways 132

Mastering For Vinyl 133
A Brief History Of Vinyl 133
How A Vinyl Record Works 134
The Vinyl Signal Chain 142
How Records Are Pressed 147
New Advances In Vinyl Technology 148
Tips For Ordering Vinyl Records 150
Takeaways 151

Mastering For CD 153
CD Basics 153
How CDs Work 154
Mastering For CD 157
Delivery Formats 162
How CDs Are Made 165
Takeaways 168

Immersive Mastering 169
The Immersive Audio Backstory 169
An Introduction To Dolby Atmos 176
Sony 360 Reality Audio 184
Takeaways 186

PART II: THE INTERVIEWS

Maor Appelbaum **191**

Eric Boulanger - The Bakery **201**

Colin Leonard - SING Mastering **209**

Pete Lyman - Infrasonic Mastering **215**

Ryan Schwabe **223**

Ian Shepherd **233**

Howie Weinberg **241**

Bob Ludwig - Gateway Mastering **247**

Doug Sax - The Mastering Lab **251**

Glossary **255**

About Bobby Owsinski **267**

    Bobby Owsinski Bibliography *267*

    Bobby Owsinski LinkedIn Learning Video Courses *268*

    Bobby Owsinski Online Courses *268*

    Bobby Owsinski's Online Connections *269*

INTRODUCTION

It's already been 23 years since the first edition of The Mastering Engineer's Handbook came out, and to say that things have changed is a major understatement. It's safe to say that there's been a continuing revolution in the mastering world, with old technologies replaced and revived, and new ones constantly evolving. Gone are the days of tape machines (for the most part), and soon even the CD might be a thing of the past (although the format continues to defy predictions and survive).

Most mastering engineers today don't feel the need to invest in "heavy iron" customized outboard gear that used to be necessary for a high-quality mastering job, since much of that particular sound can now be duplicated "in-the-box."

Even though the basic mastering tools are still the same, they've mostly moved into the world of the DAW, so even someone with the most entry-level system now has a set of powerful tools that only the top mastering pros had access to in the past. And maybe best of all, it's now possible to prep just about any kind of audio for any kind of distribution (which is what mastering really is) at home in your personal studio.

Just like everything else in music and recording, some excellent mastering tools are available to anyone with a Digital Audio Workstation. That makes the process of mastering very inexpensive compared to previous generations of musicians and audio engineers.

But just because you own a hammer doesn't mean that you know how to swing it. A lot of harm can come from misuse of these powerful tools if the process and concepts are not thoroughly understood.

And that's what this book is about.

In it, we'll take a look at how the top mastering pros perform their magic as they describe their processes in the interviews. Through this, you'll develop a strong reference point so you can either do your own mastering (and hopefully do no harm to the material, just like a doctor) or know when it's time to call a pro and properly prep the program for them to get the best results possible.

More so than any other process in audio, mastering is more than just knowing the procedure and owning the equipment. More than any other job in audio, mastering done at its highest level is about the long, hard grind of experience. It's about the cumulative knowledge gained from 12-hour days of listening to both great and terrible mixes and working on all types of music, not just the type you like.

Among the many things this book will provide is an insider's look at the process, not so much from my eyes, but from that of the legends and greats of the business.

My goal with this book is a simple one: first to show that there's a lot more to a professional mastering job than meets the eye, and with that in mind, to help you do your own mastering if that's what you want.

For those of you who have read my previous books like *The Mixing Engineer's Handbook* and *The Recording Engineer's Handbook*, you'll notice that the format of this book is similar. It's divided into two sections:

- **Part I:** *The Mechanics of Mastering* provides an overview of the history, tools, philosophy, background, and tips and tricks used by the best mastering engineers in the business.
- **Part II:** *The Interviews* is a behind-the-scenes look at the mastering world through the eyes of some of the finest (and in some cases, legendary) mastering engineers in the world.

Along with this book, you might also want to take a look at my *Mastering Audio Techniques* course at LinkedIn Learning for a more visual approach to how mastering is done.

MEET THE MASTERING ENGINEERS

Here's a list of the mastering engineers who have contributed to this book, along with some of their credits. I've tried to include not only the most notable names in the business, but also engineers who deal with specialty clients. I'll be quoting them from time to time, so I wanted to introduce them early on so you have some idea of their background when they pop up.

- Israeli-born **Maor Appelbaum** has built a diverse international client base that includes Faith No More, Limp Bizkit, Sepultura to Eric Gales and Yes. Rob Halford, Yngwie Malmsteen, Dream Theater, and many more. He's also designed several audio hardware devices and plugins.

- **Eric Boulanger** is the founder of The Bakery mastering studio, as well as a professional studio violinist. A protege of legendary mastering engineer Doug Sax, Eric has mastered GRAMMY-winning or nominated projects for Green Day, Hozier, Selena Gomez, Colbie Caillat, OneRepublic, Imagine Dragons and more.

- **Pete Lyman** is owner of Infrasonic, whose Nashville and Los Angeles locations provide services in archival, mixing, mastering, vinyl, immersive and more. Pete's mastering career spans thousands of titles over the last two decades including Grammy-Award winning and Grammy-nominated albums for Chris Stapleton, Jason Isbell, Brandi Carlile, Sturgill Simpson, John Prine, Weezer, Panic! At the Disco and more. And yes, he still cuts vinyl.

- **Ian Shepherd** has worked on thousands of CDs, DVDs and Blu-rays for all of the major record labels, TV stations and independents, including several number one singles and award-winning albums for artists like Keane, Tricky, The Royal Philharmonic Orchestra, Deep Purple, The Orb, Culture Club, Porcupine Tree, Andy Weatherall, The Las, Tentacles, New Order and King Crimson among many others. He's also developed two mastering plugins, and is the organizer of Dynamic Range Day.

- **Ryan Schwabe** is a two-time Grammy-nominated, platinum-certified mixing and mastering engineer. In 2020, he was nominated for a Grammy for Best Dance/Electronic Album for Baauer's "Planet's Mad," and in 2023 he was nominated for Best Sound Engineered Album, Non-Classical for mixing and mastering Baynk's "Adolescence." He's also owner of the music production and engineering company Xcoustic Sound, the music technology company Schwabe Digital, and co-owner of the digital record label Rare MP3s. And he's also the developer of the new Gold Clip mixing and mastering plugin.

- Atlanta-based **Colin Leonard** and his proprietary mastering technology has managed to convert many of music top mixers into fans. With credits that include Beyoncé, Justin Bieber, Jay-Z, Bad Bunny, Migos, John Legend and many more, Colin is proving that there's a new way to look at mastering that's equally effective as the traditional techniques. Along with his custom mastering service at SING Mastering, Colin is also the creator of Aria automated online mastering, the latest trend in convenient and inexpensive mastering.

- **Howie Weinberg** has 20 Grammy Awards and 76 Grammy nominations, 4 TEC awards, 2 Juno awards, 1 Mercury Prize award, over 200 gold and platinum records, an unbelievable 8,600 total credits that have accumulated 91 billion streams. His credits include some of the originators of hip hop like Kurtis Blow, Run DMC, Grandmaster Flash and Public Enemy, to legends like U2, Nirvana, Sheryl Crow, The Clash, Madonna, Alice Cooper, Aerosmith, John Mellencamp, Ozzy Osbourne – the list goes on.

I've also included edited interviews of some of the legends of mastering from previous books, since it's useful to understand that mastering really hasn't changed all that much in 50+ years.

- **Doug Sax.** Perhaps the Godfather of all mastering engineers, Doug was the first independent when he started his famous Mastering Lab in Los Angeles in 1967. Since then, he has worked his magic with such diverse talents as The Who, Pink Floyd, The Rolling Stones, the Eagles, Kenny Rogers, Barbra Streisand, Neil Diamond, Earth, Wind & Fire, Diana Krall, Dixie Chicks, Rod Stewart, Jackson Browne, and many, many more.

- **Bob Ludwig.** Bob is considered one of the giants in the mastering business, with 13 Grammy Awards and 22 other nominations coming from his more than 3,000 credits. Among his massive credit list include legendary records from Queen, U2, Sting, the Police, Janet Jackson, Mariah Carey, Guns N' Roses, Rush, Mötley Crüe, Megadeth, Metallica, David Bowie, Paul McCartney, Bruce Springsteen, the Bee Gees, Madonna, Elton John, and Daft Punk.

As a bonus, there are a number of additional interviews with mastering legends that you can access online at bobbyowsinski.com/mastering. These include:

- **Bernie Grundman.** One of the most widely respected names in the recording industry, Bernie has mastered literally hundreds of platinum and gold albums, including some of the most successful landmark recordings of all time, such as Michael Jackson's *Thriller*, Steely Dan's *Aja*, and Carole King's *Tapestry*.

- **Bob Katz.** Co-owner of Orlando-based Digital Domain, Bob specializes in mastering audiophile recordings of acoustic music, from folk to classical. The former technical director of the widely acclaimed Chesky Records, Bob's recordings have received disc of the month in Stereophile and other magazines numerous times, and his recording of "Portraits of Cuba" by Paquito D'Rivera, won the 1997 Grammy for Best Latin-Jazz Recording. Bob's mastering clients include major labels EMI, WEA-Latina, BMG, and Sony Classical, as well as numerous independent labels.

- **Greg Calbi.** One of the owners of Sterling Sound in New York City, Greg's credits include Bob Dylan, John Lennon, U2, David Bowie, Paul Simon, Paul McCartney, Blues Traveler, and Sarah McLachlan, among many, many others.

- **Glenn Meadows.** Glenn is a Nashville-based two-time Grammy winner and a multi–TEC award nominee who has worked on scores of gold and platinum records for a diverse array of artists, including Shania Twain, LeAnn Rimes, Randy Travis, Delbert McClinton, and Reba McEntire, as well as for multi-platinum producers such as Tony Brown, Jimmy Bowen, and Mutt Lange.

- **Bob Ohlhsson.** After cutting his first number one record (Stevie Wonder's "Uptight") at age 18, Bob worked on an amazing 80 top ten records while working for Motown in Detroit. Now located in Nashville, Bob's unique view of the technology world and his insightful account of the history of the industry makes for a truly fascinating read.

- **Gene Grimaldi.** Gene is the chief engineer at Oasis Mastering in Los Angeles, and has a list of blockbuster clients that include Lady Gaga, Jennifer Lopez, Carly Rae Jepsen, Ellie Goulding, Nicki Minaj, and many more.

- **David Glasser.** David is the founder and chief engineer of Airshow Mastering in Boulder, Colorado, and Takoma Park, Maryland, and has worked for some 80 Grammy nominees. He's also an expert in catalog restoration, having worked on releases by Smithsonian Folkways Recordings and the Grateful Dead, among many others.

- **Dave Collins.** Operating out of his own Dave Collins Mastering studios in Hollywood, Dave has mastered projects for Sting, Madonna, Bruce Springsteen, and Soundgarden, among many others.

- **Eddy Schreyer.** Noted veteran engineer Eddy opened Oasis Mastering in 1996 after mastering stints at Capitol, MCA, and Future Disc. With a list of chart topping clients that span the various musical genres such as Babyface, Eric Clapton, Christina Aguilera, Fiona Apple, Hootie and the Blowfish, Offspring, Korn, Dave Hollister, Pennywise, and Exhibit, Eddy's work is heard and respected world-wide.

- While you probably won't have access to the gear, playback systems, and rooms that the above engineers have, that's okay because a great mastering job can be at your fingertips if you follow their advice and examples, and use the greatest tool you have available—your ears.

PART I
THE MECHANICS OF MASTERING

THE ESSENCE OF MASTERING

The term "mastering" is either completely misunderstood or shrouded in mystery, but the process is really pretty simple. Technically speaking, mastering is the intermediate step between mixing the audio and having it replicated or distributed. Up until recently, we would define it as follows:

> ***Mastering is the process of turning a collection of songs into an album by making them sound like they belong together in tone, volume, and timing (spacing between songs).***

That was the old way to explain mastering when the album was king. Since we live in a singles world today, the definition has to be tweaked for our current production flow. Let's use this definition instead.

> ***Mastering is the process of fine-tuning the level, frequency balance, and metadata of a track in preparation for distribution.***

That first definition isn't obsolete though, since albums are still around (and probably always will be), but the fact of the matter is that individual songs are always played in a collection at some point. The collection can be a physical album like a CD or vinyl record or, more commonly, a playlist where the song is played before or after someone else's track on the radio, or on an online distribution service like Spotify or Apple Music. Of course, you want all of your songs to sound at least as good as the others that you listen to or the ones played before or after them.

> *I think that mastering is a way of maximizing music to make it more effective for the listener as well as maybe maximizing it in a competitive way for the industry. It's the final creative step and the last chance to do any modifications that might take the song to the next level.*
> —Mastering legend Bernie Grundman

So loosely speaking, that's what mastering is. **Here's what mastering is not**—it's not a tool or a plugin that automatically masters a song with little or no effort from the operator. All too often people have the misconception that mastering is only about EQing a track to make it sound bigger, but it's really more of an art form that relies on an individual's skill, experience with various genres of music, and good taste. In fact, it's often said that 95 percent of all mastering is in the ears, and not the tools.

Mastering is about having a conversation with a person. It's about me understanding the artist's intent and trying to help them realize it to make their music the best it can be. It's also about understanding that some songs are meant to be quiet, others loud, some meant to be angry even though they're quiet, or fun even though they're loud. Mastering is about grasping the emotion behind the music, and AI tools just aren't there yet—and I'm not sure they ever will be.

—Ian Shepherd

The more work you do, the better you become at handling different situations, because mastering is a numbers game—the more you do, the more you learn.

—Maor Appelbaum

While the tools for audio mastering do require more precision than in other audio operations, the bottom line is that this is an area of audio where experience really does matter.

WHY MASTER ANYWAY?

Mastering should be considered the final step in the creative process, as this is the last chance to polish and fix a project. Not all projects need mastering, especially if they're not destined to be heard by the public, but here are a few instances when mastering can help:

- If you have a song that sounds pretty good by itself but doesn't sound as loud as other songs.
- If you have a song that sounds pretty good by itself but sounds too bright or dull next to other songs.
- If you have a song that sounds pretty good by itself but sounds too bottom heavy or bottom light compared to other songs.

A well-mastered project simply sounds better *if the mastering is done well* (that's the key phrase, of course). It sounds complete, polished, and finished. The project that might have sounded like a demo before now sounds like a "record" because:

- Judicious amounts of EQ and compression are added to make the project sound bigger, fuller, richer, and louder.
- The levels on each song of the album (if there is one) are adjusted so they all have the same apparent level or have the same level as other professionally mastered songs in the same genre.
- The fades aren fixed, if needed, so that they're smooth.
- Any distorted parts or glitches are edited out.
- All the songs of an album blend together into a cohesive unit.
- In the case of mastering for CD or vinyl, the spreads (the time between each song) are inserted so the songs flow seamlessly together.

- The songs destined for a CD or vinyl record are sequenced so they fall in the correct order.
- ISRC codes and the proper metadata are inserted into each track.
- A backup clone is created and stored in case anything happens to the master.
- Any shipping or uploading to the desired replication facility is taken care of.

As you can see, there's a lot more to mastering than meets the eye and ear when you really get into it. To better understand mastering, let's see how it has evolved over the years.

FROM VINYL, TO CDS, TO MP3S, AND TO STREAMING

Until 1948, there was no distinction between different types of audio engineers because everything was recorded directly onto 10-inch vinyl records that played at 78 rpm. In 1948, however, the era of the mastering engineer began when Ampex introduced its first commercial magnetic tape recorder.

Since most recording at the time began using magnetic tape, a transfer had to be made to a vinyl master for delivery to the pressing plant to press records, hence the first incarnation of the "mastering engineer" was born. There was no concept of the process that we now consider to be mastering at the time though, so they were called a "transfer engineer." (see Figure 1.1).

Figure 1.1: A disc-cutting lathe
© 2024 Bobby Owsinski

This transfer process was highly challenging because the level applied to the master vinyl lacquer when cutting the grooves was so crucial. Too low a level and you get a noisy disc, but hit it too hard and you destroy the disc and maybe the expensive ($15,000 in '50s and '60s dollars) cutting stylus of the lathe too (see Figure 1.2).

Figure 1.2: A disc-cutting stylus
© 2024 Bobby Owsinski

In 1955, Ampex released tape machines that had a new feature called Selective Synchronous Recording, or Sel Sync, which gave the multitrack recorder the ability to overdub new tracks, thus changing the recording industry forever.

At this point there became a real distinction between the recording and mastering engineer, since the jobs now differed so greatly, although many were trained at both jobs (the EMI training program at Abbey Road made mastering the last job before you became a full engineer).

In 1957, the stereo vinyl record became commercially available and really pushed the industry to what many say was the best-sounding audio ever. Mastering engineers, who were now known as "cutters," found ways to make the discs louder (and as a result less noisy) by applying equalization and compression.

Producers and artists began to take notice that certain records would actually sound louder on the radio, and if it played louder, then the listeners usually thought it sounded better (although they were speculating instead of using any scientific data at the time), and maybe the disc sold better as a result. Hence, a new breed of mastering engineer was born—this one with some creative control and ability to influence the final sound of a record, rather than just simply transferring the audio from medium to medium.

An interesting distinction between American and British mastering engineers developed though. In the U.S., mastering was —and still is— considered the final step in the creation of an album, while in the UK they look at it as the first step in manufacturing. As a result, American mastering engineers tended to have much more creative leeway in what they were allowed to do to the audio than British engineers.

With the introduction of the CD in 1982, the cutting engineer—now finally known as a "mastering engineer"—was forced into the digital age. Because of the limitations of digital storage at the time, they were forced to use a special processor and a modified video tape recorder called a Sony 1630 (see Figure 1.3) to deliver the digital CD master to the replicator. Even though the storage was digital, they were still utilizing many of the analog tools from the vinyl past for EQ and compression.

The 1989 introduction of the Sonic Solutions digital audio workstation with "pre-mastering software" provided a CD master instead of a bulky 1630 tape cartridge (see Figure 1.4). Now mastering began to evolve into the digital state as we know it today.

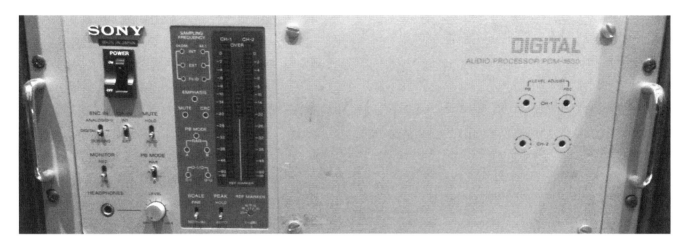

Figure 1.3: A Sony 1630
© 2024 Bobby Owsinski

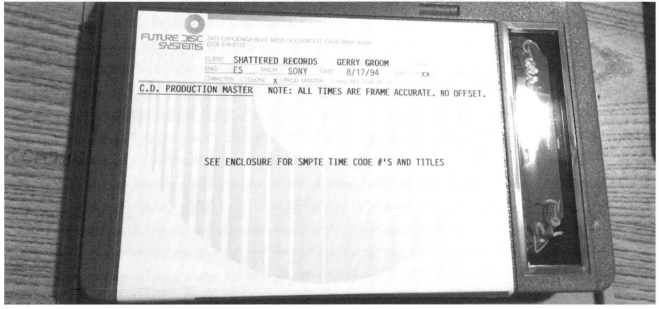

Figure 1.4: A tape cartridge used in a 1630
© 2024 Bobby Owsinski

In the first half of 1995, MP3s began to spread on the Internet, and their small file size set about a revolution in the music industry that continues to this day. This meant that the mastering engineer had to become well-versed in how to get the most from this format, something it took many mastering engineers time to get the hang of.

In 1999, 5.1 surround sound and high-resolution audio took the mastering engineer into new, uncharted but highly creative territory, just like immersive mastering today. And by 2002, almost all mastering engineers were well acquainted with the computer, since virtually every single project was edited and manipulated with digital audio workstation software.

Nowadays a majority of engineers are firmly in the box unless given the increasingly rare 1/2- or 1/4-inch tape to master.

Today's mastering engineer doesn't practice the black art of disc cutting as much as was once required, but they are no less the wizard, continuing to shape and mold a project like never before.

THE DIFFERENCE BETWEEN YOU AND A PRO

There are a lot of reasons why a commercial mastering facility usually produces a better product than mastering at home or through an online service. If we break it down, a mastering pro usually has three things over the home studio.

The Gear: A real pro mastering studio has many things available that you probably won't find in a simple home or small studio DAW room, such as high-end A/D and D/A converters with a superior signal path, a great-sounding listening environment, and an exceptional monitoring system (see Figure 1.5).

Figure 1.5: The Tannoy monitoring system at Oasis Mastering
© 2024 Bobby Owsinski

> *The reason people come to a mastering engineer is to gain that mastering engineer's anchor into what they hear and how they hear it and the ability to get that stuff sounding right to the outside world.*
> —Mastering legend Glenn Meadows

The monitor systems of these facilities often costs far more than many entire home studios (and even more than some homes, for that matter). Cost isn't the point here, but quality is, since you can rarely hear what you need to on the nearfield monitors that most recording studios use in order to make the fine adjustments that are needed in a great mastering job. The vast majority of monitors, and the rooms in which they reside, are simply not precise enough.

The Ears: The mastering engineer is the real key to the process. This is all they do day in and day out. They have "big ears" because they master for at least eight hours every day and know their monitors better than you know your favorite pair of sneakers. Plus, their reference point of what constitutes a good-sounding mix is finely honed thanks to countless hours and hours of listening to best- and worst-sounding mixes of each genre of music.

> *Most people need a mastering engineer to bring a certain amount of objectivity to their mix, plus a certain amount of experience. If you (the mastering engineer) have been in the business a while, you've listened to a lot of material, and you've probably heard what really great recordings of any type of music sound like, so in your mind you immediately compare it to the best ones you've ever heard. You know, the ones that really got you excited and created the kind of effect that producers are looking for. If it doesn't meet that ideal, you try to manipulate the sound in such a way as to make it as exciting and effective a musical experience as you've ever had with that kind of music.*
> —Bernie Grundman

A Backup: I don't know who said it, but this phrase rings true: "The difference between a pro and an amateur is that a pro always has a backup." Good advice for any part of recording, but especially for mastering. You wouldn't believe the number of times masters get lost, even when major record labels are involved. This is one thing that you can do just as well as a pro can with no trouble at all!

Finally, if mastering was so easy, don't you think that every big-time engineer or producer (or record label, for that matter) would do it themselves? They don't, and the major mastering houses are busier than ever, which tells you something.

> *I tell the interns who come through here all the time: if you want the record to sound consistent, focus on the vocal. In 95% of music, the vocal is the most important element. It needs to be consistent across the record. That's what people are really listening to.*
> —Pete Lyman

There's Always Room for DIY

While the above section may seem like I'm trying to discourage you from doing your own mastering, that's not the case. What I'm trying to do is give you a reference point of how the pros operate and why they're so successful. From there you can determine whether you're better served by doing it yourself or using a pro.

But the reason that you're reading this book is because you want to learn about all the tricks, techniques, and nuances of a major mastering engineer, right? Some mastering situations don't need a professional's touch, while other times budgets are so tight that there's just no money left over for a mastering pro no matter how much you'd like to use one.

> *As far as the person who might be trying to learn how to do his own mastering, or understand mastering in general, the main thing is that all you need is one experience of hearing somebody else master something. Your one experience at having it sound so incredibly different makes you then realize just how intricate mastering can be and just how much you could add to or subtract from a final mix.*
>
> —Greg Calbi - Grammy-winning mastering engineer at Sterling Sound

> *Just owning a Pro Tools system does not make you a mastering engineer.*
>
> —Mastering Legend Doug Sax

Read on, and you'll discover the hows and whys of mastering in detail.

TAKEAWAYS

- Mastering is the process of fine-tuning the level, frequency balance, and metadata of a track to prepare it for distribution.

- Mastering is not a tool or a plugin that automatically masters a song with little or no effort from the operator.

- Mastering should be considered the final step in the creative process, as it offers the last opportunity to polish and fix a project.

- A professional mastering engineer has a better listening environment, better monitors, higher quality gear and more experience than almost all musicians and engineers.

DIGITAL AUDIO BASICS

Now is a good time for a brief review the basics of digital audio. While you may be familiar with the sample rate and word length already, there are often many questions about the differences between file formats, such as AIFF and WAV, so we'll try to take care of them straight away.

Sample rate and word length determine the quality of a digital audio signal. To understand how that happens, a brief discussion is necessary. Remember, this is just an overview and only gives you the general concepts of digital audio. If you really want a deeper dive into digital audio, refer to a book like Principles of Digital Audio by Ken Pohlmann that thoroughly covers the subject.

SAMPLE RATE

Sample rate is one of the key factors when it comes to the quality of a digital audio signal. The analog audio waveform amplitude is measured by the analog-to-digital converter (more on this device in Chapter 4, "Monitoring for Mastering") at discrete points in time, and this is called sampling. The more samples that are taken of the analog waveform per second, the better the digital representation of the waveform is, which results in a greater frequency response of the signal (see Figure 2.1).

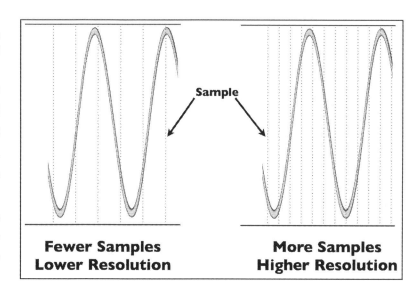

Figure 2.1: Sample rate
© 2024 Bobby Owsinski

For example, if we were to use a sampling rate of 48,000 times a second (or 48kHz), that would present us with a frequency response of 24kHz, or half the sample rate. This is due to the Nyquist Theorem, a fundamental principle of digital audio which states that *the sample rate has to be twice as high as the highest frequency you wish to record,* otherwise, digital artifacts called aliasing will be added to the signal. A low-pass filter is applied to limit the bandwidth to half the sampling rate.

A sample rate of 96kHz provides a better digital representation of the waveform because it takes more samples, and yields a usable audio bandwidth of about 48kHz. A 192kHz sample rate provides a bandwidth of 96kHz. Some recordings now even go as high as 384kHz, giving you a potential bandwidth of 192kHz.

While it's true that we can't hear above 20kHz on even a good day, the increased frequency response that's available as a result of the high sampling rate allows the use of a less intrusive filter. This results in a better sounding digital signal, which is why we strive to use higher sample rates if possible.

TIP: The higher the sampling rate, the better the representation of the analog signal and the greater the audio bandwidth will be, which means it sounds better!

Although a higher sample rate provides a better representation of the analog signal, some people may not hear the difference due due to factors like the speakers, the listening environment, the signal path, the type of music, or how it was mixed. Couple that with the fact that higher sample rates require more powerful computers, fewer tracks and plugins are available, and some plugins won't work at some of the very high sample rates. For these reasons you can see why sometimes a lower sample rate can be a better decision when it comes to recording.

That said, 96kHz has become the new standard for music recording, especially since Apple's iTunes and then Apple Digital Masters program encourages delivery of high-resolution files at that rate.

The downside of a higher sample rate is that it consumes more digital storage space, with 96kHz taking up twice as much as 48k, and 192k taking up twice as much again as 96k. That's no longer much of a problem though, as hard-drive disk, SSD, or even flash-drive storage is massive compared to the needs of a typical song.

TIP: It's always best to mix at the highest resolution possible both for archival purposes and because a high-resolution master makes for a better-sounding lower-resolution file. This applies even if the ultimate delivery medium is to be a lower-resolution CD or MP3.

That said, a mastering engineer must work at certain sampling rates to deliver a product for a particular distribution medium.

TYPICAL SAMPLE RATES	COMMENTS	CAVEATS
44.1kHz	The CD sample rate	Fewer CDs are being made, so the minor advantage of recording using the similar sample rate is lost.
48kHz	Standard for film and TV	Lowest recommended sample rate.
96kHz	High-resolution standard	Most pro records are recorded at 96kHz. The recommended master delivery rate for Apple Music. However, it takes up twice the storage space of 48kHz.
192kHz	Audiophile standard	Only half the channels and plugins of 96kHz on some DAWs. However, many plugins don't operate, and it takes up twice the storage space of 96kHz.

BIT DEPTH

Bit depth refers to the length of a digital word, and it's another factor determining audio quality. Like sample rate, more is generally better.

The more bits in a digital word, the greater the dynamic range, which once again means the final audio sounds more realistic. Each extra bit that's used provides 6dB more dynamic range available. Therefore, 16 bits yields a maximum dynamic range of 96dB, 20 bits provides 120dB, and 24 bits offers a theoretical maximum of 144dB. From this you can see that a high-resolution 96kHz/24-bit (commonly abbreviated 96/24) format is far closer to sonic realism than the CD standard of 44.1kHz/16-bit.

Today almost all professional recording is done at 24 bits, as there's virtually no advantage to using a lower bit depth.

While at one time hard drive space or bandwidth was at a premium, that's no longer the case. That said, both CD and MP3 formats require 16 bits, but online streaming programs like Apple Digital Masters now encourage 24-bit delivery regardless of the sample rate.

TIP: The longer the word length (the more bits), the greater the dynamic range and therefore the closer to "real" the sound can be.

Even though bit depths like 32-bit, 32-bit float, and 64-bit float potentially provide much higher quality (especially with plugins), it's unlikely that you've not recorded in these higher resolutions, so consider them for a future delivery application.

TIP: Never export to higher resolution than the one your project started at, since it won't improve the quality, and and will result in a significantly larger file. For instance, if your project started at 16-bit, selecting 24-bit or higher offers you no benefit.

STANDARD AUDIO FILE FORMATS

Several audio file formats are used today on most digital audio workstations. A file format specifies how the digital word is encoded in a digital storage medium. Some file formats are universal and others are proprietary, while some, such as JPEG and PNG files for graphics, are very specific to the type of information they store. There are also specific types of file formats used for audio.

LPCM (Linear Pulse Code Modulation) is the process of sampling an analog waveform and converting it into digital bits that are represented by binary digits (1s and 0s) of the sample values. When LPCM audio is transmitted, each "1" is represented by a positive voltage pulse, and each "0" is represented by the absence of a pulse (see Figure 2.2). LPCM is the most common method of storing and transmitting uncompressed digital audio.

Since it's a generic format, it can be read by most audio applications—similar to the way a plain text file can be read by any word-processing program. LPCM is used by digital audio workstations in a file formats like AIFF, BWF, or WAV files.

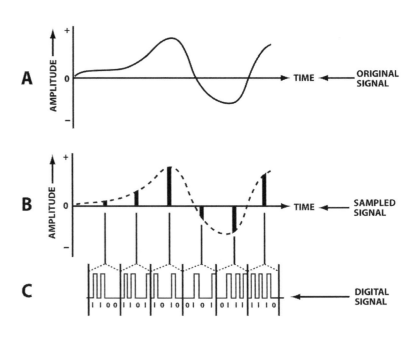

Figure 2.2 : Linear PCM
© 2024 Bobby Owsinski

AIFF (Audio Interchange File Format) is used to store LPCM digital audio data. It supports a variety of bit resolutions, sample rates, and channels of audio. The format was developed Apple and is the standard audio format for Macintosh computers, although it can be read by most computer workstations. AIFF files generally end with .aif.

WAV (Waveform Audio) is another file format for storing LPCM digital audio data. Created by Microsoft and IBM, WAV was one of the first audio file types developed for the PC. Wave files are indicated by a .wav suffix in the file name and are often spelled wav (instead of wave) in writing. The .wav file format supports a variety of bit resolutions, sample rates, and channels of audio.

BWF (Broadcast Wave) is a special version of the standard WAV audio file format developed by the European Broadcast Union in 1996. BWFs contain an extra chunk of data, known as the *broadcast extension chunk*, that contains information on the author, title, origination, date, time, and so on of the audio content.

Perhaps the most significant aspect of BWFs is their time stamping feature, which allows files to be transferred from one DAW application to another and easily aligned to their correct position on a timeline or an edit decision list.

Today's there's really no operational difference between AIFF and WAV files. In the past, you'd use an AIFF audio file if you were on a Mac and a WAV file if you were on a PC, but both platforms now happily read either one without any difficulty.

CAF (Core Audio Format). Some DAWs can export a new file format developed by Apple around its Core Audio technology for use with operating systems 10.4 and higher. CAFs are designed to overcome some of the limitations of the older WAV and AIFF file containers, such as the limit on file size and tracks. CAF files don't have the 4GB limit of the other formats, and can theoretically hold a file that is hundreds of years long (that's a massive file!).

The format is also able to hold practically any type of audio data and metadata, any number of audio channels, and auxiliary information such as text annotations, markers, channel layouts, and other DAW data.

An interesting feature of the CAF file format is that you can append new audio data the end of the file, making it ideal as an archive format.

All Apple software products, including Logic Pro, GarageBand, and QuickTime Player, as well as T-RackS, now support CAF files and can open them directly. If you need CAF files to be played on other systems, convert them to a WAV or MP3 file using a utility such as Format Factory.

DATA COMPRESSION

Linear PCM files are large and can sometimes be painfully slow to upload and download if you don't have a high-speed Internet connection. As a result, data compression was introduced to retain some level of sonic integrity (how much is in the ear of the beholder) while making an audio file more easily transportable.

Data compression isn't at all like the audio compression that we're going to be talking about later in the book. Data compression reduces the amount of physical storage space and memory required to store an audio file, and therefore reduces the time required to transfer a file.

File formats that use data compression include MP3, AAC, FLAC, Dolby Digital, DTS, and many more. Check out Chapter 10 for more on the different types of data compression.

TAKEAWAYS

- The higher the sampling rate, the better the representation of the analog signal and the greater the audio bandwidth, which means it sounds better!

- 44.1kHz is the standard sampling rate for CD, 48kHz for film and television, and 96kHz for high-resolution audio.

- The longer the word length (the more bits), the greater the dynamic range and the closer the sound is to "real."

- LPCM is used by audio workstations and is represented by file format by AIFF, BWF, or WAV files.

- Data compression reduces the physical storage space and memory required to store an audio file, and therefore reduces the time required to transfer it.

3

PREPPING FOR MASTERING

In order for a mastering session to run smoothly, sound great, and save you money, some prep work is required beforehand. Even if you're doing your own mastering or using an online service, these tips can help improve your end product.

MIXING FOR MASTERING

Nothing is as exasperating to all involved in the mastering process as not knowing which mix is the correct one or forgetting the file name. Here are some tips to get your tracks mastering-ready.

- **Don't over-EQ when mixing.** A mix is over-EQ'd when it has big spikes in its frequency response as a result of trying to make one or more instruments sit better in the mix. This can make your mix painfully bright, or give it a huge unnatural-sounding bottom. In general, mastering engineers can do a better job for you if your mix is on the dull side rather than too bright. Likewise, it's better to be light on the bottom end than to have too much.

- **Don't over-compress when mixing.** Over-compression means that you've added so much mix bus compression that the mix is robbed of all its dynamic range and feels lifeless as a result. You can tell that a mix has been over-compressed not only by its sound, but by the way its waveform is flat-lined on the DAW timeline.

    You might as well not even master if you've squashed it too much already. Hyper-compression (see Chapter 6, "Mastering Techniques") deprives the mastering engineer of one of his or her major abilities to help your project. Squash it for your friends, squash it for your clients, but leave some dynamics in the song so the mastering engineer is better able to do their job. In general, it's best to compress and control levels on an individual-track basis and not as much on the stereo bus, except to prevent digital overs.

- **Matching levels between songs isn't too important.** Just make your mixes sound great, because matching levels between songs is one of the reasons you master in the first place.

- **Getting hot mix levels is not important.** Today dynamic range is much more important than hot levels (see Chapter 8).

- **Be careful of your fades and trims.** Trimming the heads and tails of a track too tightly might cut off a reverb trail or an essential attack or breath. Leave a little room and perfect it during mastering, where you will probably hear things better.

- **Have the right documentation.** See the next section.

- **Print the highest-resolution mixes possible by staying at the same resolution as the tracks were recorded at.** Print the highest-resolution mixes possible by staying at the same resolution as the tracks were recorded at. In other words, if the tracks were cut at a sample rate of 96kHz/24-bit, that's the resolution your mix should be. If it's at 44.1kHz/24-bit, that's the resolution the mix should be.

- **Don't add dither.** Adding dither to your mix actually reduces the resolution. Leave that for the mastering process (see more about dither in Chapter 7).

- **Alternate mixes can be your friend.** A vocal up/down or instrument-only mix can be a lifesaver when mastering. Things that aren't apparent while mixing sometimes jump right out during mastering, and having an alternative mix around can sometimes provide a quick fix and keep you from having to remix. Make sure you document all alternative mixes properly though.

- **Reference your mixes in mono when mixing.** It can be a shock when the mastering engineer checks for mono compatibility, and the lead singer or guitar solo disappears because something is out of phase. Even though this was more of a problem in the days of vinyl and AM radio, it's still an important point because many so-called stereo sources (such as television and some smart speakers) are either pseudo-stereo or only heard in stereo some of the time. Check it and fix it before the mastering session.

- **Know your song sequence.** Song sequencing takes a lot of thought in order to make an album flow, so you don't want to leave that undecided until the mastering session. If you're cutting vinyl, remember that you need two sequences—one for each side of the disc. Remember, the masters can't be completed without the sequence. Also, unlike in a DAW, cutting vinyl is a one-shot deal with no chance to undo. It'll cost you money every time you change your mind.

- **Have your songs timed out.** This is important if you're going to be making a CD, cassette, or a vinyl record. First, you want to make sure that your project can easily fit on a CD, if that's your release format. Most CDs have a total storage time of just under 80 minutes, so that time shouldn't be much of a problem unless it's a concert or a double album. When mastering for vinyl, cumulative time is important because the mastering engineer must know the total time per side before the cutting begins. Due to the physical limitations of the disc, you're limited to a maximum of about 25 minutes per side if you want the record to be loud and somewhat noise-free.

These days, I think I see too much harmonic saturation and unnecessary "monkey business" going on in mixes. Don't get me wrong, I love grit, distortion, and overdrive—I have tons of transformer options here, and I'm a fan of that stuff. But I think people tend to overuse it. Once that distortion is baked into a mix, you can't get rid of it, and if the project is going to vinyl, that could cause problems.
—Pete Lyman

There's just too much emphasis on making everything loud. It doesn't leave me as much room to work with, and it takes away from the overall quality and musicality of the sound.
—Ian Shepherd

Mastering Session Documentation

You'll make it easier on yourself and your mastering engineer (even it that's you) if everything is well documented, and you'll save yourself some money, too. Here's what to include:

- **The title of the album and songs.** Use the final titles, not shortened working versions, with the exact spelling that will appear on the final product.
- **The metadata, including the artist, label, and publishing information,** especially if you want the mastering facility to make MP3 files for you.
- **Any flaws, digital errors, distortion, bad edits, fades,** or anything out of the ordinary that a file might have.
- **Any FTP or shipping instructions for sending your master to a replicator.**
- **Any ISRC and UPC codes.** We'll go over both in depth in Chapter 6.
- **Properly ID'd files.** Make sure that all files are properly titled for easy identification (especially if you're not there during the session), including alternative mixes.
- **A mastering reference.** Providing the mastering engineer with a commercially released CD that has a sound you really like may provide a reference point. This is also important when using online mastering services or some mastering plugins.

Having the right documentation can make your mastering session go a lot faster, which can be important, especially when you're trying to make a release date. All it takes is a little bit of forethought and preparation.

Why Alternative Mixes Can Be Essential During Mastering

Even though mixes in a DAW can be almost instantly recalled and changed, most mixers still print alternate mixes to make ultra-quick fixes during mastering possible.

While alternate mixes with the vocal raised or lowered by 1dB used to be the norm, today's mixers find that three types of alternate mixes can handle most fixes:

- **The instrumental mix.** This is often used to clean up objectionable lyrics on a song by editing in a small piece over the final mix. That way, the mix sounds a lot better than if a word is bleeped out with an audio tone. It's also sometimes used for licensing to television shows.

- **The voice-only mix.** By using a combination of the instrumental mix with the acapella mix, it's possible to adjust the level of a word that might be too loud or masked.

- **The TV mix.** The TV mix has everything but the lead vocal, allowing the artist or band to perform live on television against a prerecorded background. Sometimes it's used instead of an instrumental mix.

Although editing may be an overlooked skill of the mastering engineer, it can come in handy when alternate mixes are available. Even though mix fixes in a DAW can be fast, sometimes using the alternate mixes to make a fix can be even faster. If you're on the fence about the level or EQ of a certain instrument in the mix, print a couple of options.

The philosophy of getting it right the first time should carry through every step of the process—from the first note played to delivering the album. If the sound coming out of the speakers isn't what you want it to be, you need to address it right then.

—Eric Boulanger

TAKEAWAYS

- Even if you're mastering your own material, prepping your mix for mastering will ensure that your mastering session run more smoothly.

- The three types of alternative mixes most helpful to the master engineer are the instrumental mix, voice-only mix, and TV mix.

- The TV mix contains everything but the lead vocal, allowing the artist or band to perform live on television while singing live against a prerecorded background.

4

MONITORING FOR MASTERING

The first time I had a song mastered I remember being amazed at the initial playback. I was hearing things that I never heard before (most of them not too flattering) and areas of the mix that suddenly jumped out as wrong. I could hear reverb trails and fades that were cut off, vocals that were too harsh, and an inflated low end that I had no idea was there. In other words, I could suddenly hear all the flaws that weren't even on my listening radar before.

That's exactly what jumps out to you in a purpose-built mastering facility - the clarity and truth about what you're hearing. It's clinical, precise, and way beyond what you're used to hearing even in the best recording studio.

That's because the heart and soul of the mastering signal chain is the loudspeakers. More than any one device, these are the main link of the mastering engineer to both the reference point of the outside world and the possible deficiencies of the source material. More care is taken when it comes to choosing the monitoring system than just about any other piece of gear in the mastering studio.

> *Probably the one biggest and most important piece of equipment that a mastering engineer can have is their monitor, and they has to understand that monitor and really know when it's where it should be. If you know the monitor and you've lived with it for a long time, then you're probably going to be able to make good recordings.*
>
> —Bernie Grundman

THE ACOUSTIC ENVIRONMENT

Having the finest reproduction equipment is useless unless the acoustic environment in which they're placed is optimized. Because of this, more time, attention, and expense is initially spent on the acoustic space than on virtually any other aspect in a high-end mastering facility.

> *I think a lot of people have heard about the effort we've gone through to make our room as acoustically perfect as possible. Many times people come into the room and go, "Oh my God!" or something like that.*
>
> —Bob Ludwig

That said, when it comes to mastering your own material in your personal studio, the single greatest impediment to doing an acceptable job can be your monitoring environment, which is why it's so important to try to improve it.

I know you most likely can't afford George Augspurger to trick out your garage (he's considered the father of modern studio acoustic design), and those ATC monitors are still about $30,000 out of reach. Don't worry, there's still hope.

LET'S FIX YOUR LISTENING AREA

The listening environment is probably the most overlooked part of most home studios. While it's easy to spend a lot of money trying to improve your listening area, here are a few zero-cost placement tips that in some cases can really make a difference.

- **Avoid placing speakers up against a wall.** The farther away you can get from the wall, the smoother the monitor speaker response will be, especially with the frequencies below 100Hz.

- **Avoid the corners of the room.** A corner reinforces the low end even more than when placed against a wall. The worst case is if only one speaker is in the corner, which will cause the low-end response of your system to be lopsided.

- **Avoid placing one speaker closer to a wall than the other.** If one speaker is closer to a side wall than the other, you'll get a totally different frequency response between the two because the reflections from the wall are different than on the other side. It's best to set up equidistant from each wall if possible.

- **Avoid different types of wall absorption on each side of the room.** If one side of the room uses a wall material that's soft and absorbent while the other is hard and reflective, you'll have an unbalanced stereo image because one side will be brighter than the other. Try to make the walls on each side of the speakers the same in terms of the amount of absorption that is used.

The above tips aren't a cure-all for big acoustic problems, and they won't do anything for your isolation (it still takes big bucks for that), but you'd be surprised how much better things can sound by just moving the gear around the room.

Find out additional tips about how to improve your listening environment in my book, *The Studio Builder's Handbook* (Alfred Music) or by watching my *Music Studio Setup and Acoustics* video series on linkedinlearning.com.

THE MONITORS

Pro mastering facilities will always choose a mastering monitor with a wide and flat frequency response. Wide frequency response is especially important on the bottom end of the frequency spectrum, which means that a relatively large monitor is required, perhaps with an additional subwoofer as well.

Because of this requirement, many of the common monitors used in recording and mixing, especially nearfields, will not provide the frequency response needed for a typical mastering scenario.

Smooth frequency response is important for a number of reasons. First, an inaccurate response will result in inaccurate equalization in order to compensate. It will also probably mean you'll overuse the EQ in an unconscious attempt to overcome the deficiencies of the monitors themselves.

Let's say that your monitors have a dip at 2kHz (not uncommon, since that's about the crossover point of most two-way nearfield monitors). While recording and mixing, you boost 2kHz to compensate for what you're not hearing. Now, it might sound okay on these monitors during your mastering process, but if you play it back on another set of speakers, you might find that the midrange is suddenly tearing your head off.

Now let's bring the environment into the equation, which compounds the problem. Let's say that between the monitors you're listening to (like a typical two-way system with a 6- or 8-inch woofer) and your room, you're not hearing anything below 60Hz or so. To compensate, you add +8dB of 60Hz so it sounds the way you think it should sound. If you master on the same monitors in the same environment, you'll never realize that when you play the song outside of your studio, it will be a big booming mess.

When it comes to mastering, large monitors with a lot of power behind them are not for loud playback, but for clean and detailed, distortion-free level. These monitors never sound "loud" the way we're used to, they just sound bigger and clearer so they reveal every nuance of the music.

Although the selection of monitors is a very subjective and personal issue (just like in recording), some brand names repeatedly pop up in major mastering houses. These include Tannoy, B&W, PMC, Lipinski, ATC, and Duntech.

> *One reason I've always tried to get the very best speaker I can is I've found that when something sounds really right on an accurate speaker, it tends to sound right on a wide variety of speakers.*
> —Bob Ludwig

> *It's not that we're going for the biggest or the most powerful sound; we're going for neutral because we really want to hear how one tune compares to the other in an album. We want to hear what we're doing when we add just a half dB at 5k or 10k. A lot of speakers nowadays have a lot of coloration and they're kind of fun to listen to, but boy, it's hard to hear those subtle little differences.*
> —Bernie Grundman

Basic Monitor Setup

Too often, musicians and engineers haphazardly set up their monitors, and this is a leading cause of mixing and mastering problems later on down the line. How the monitors are placed can make an enormous difference in the frequency balance and stereo field and should be addressed before you get into any serious listening.

Here are a few things to experiment with before you settle on the exact placement.

- **Check the distance between the monitors.** If the monitors are too close together, the stereo field will be smeared with no clear spatial definition. If the monitors are too far apart, the focal point or "sweet spot" will be too far behind you, and you'll hear the left or the right side distinctly, but not both together as one. A rule of thumb is that the speakers should be as far apart as the distance from the listening position. That is, if you're 4 feet away from the monitors, then start by moving them 4 feet apart so that you make an equilateral triangle between you and the two monitors (see Figure 4.1). A simple tape measure will work fine to get it close. You can adjust them either in or out from there.

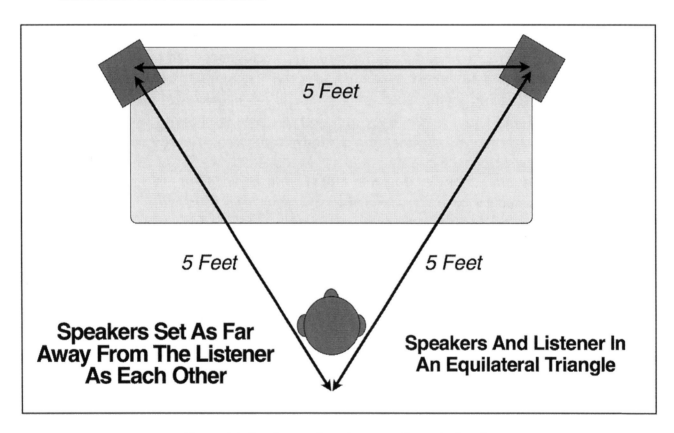

Figure 4.1: Set the monitors in an equilateral triangle
© 2024 Bobby Owsinski

That being said, it's been found that 67-1/2 inches from tweeter to tweeter at the distance of a console meter bridge seems to be an optimum distance between speakers, and focuses the speakers just behind your head (which is exactly what you want).

- **A really quick setup that we used to use in the days of consoles is to open your arms as wide as possible to each side and place the monitors at the tips of the fingers of each hand.** This seemed to work well because of the built-in depth that the console would provide, but it doesn't really apply in these days of workstations, where the monitors are a lot closer to you than ever. If that's the case, go back to the equilateral triangle outlined above.

- **Check the angle of the monitors.** Improper angling will also smear the stereo field, which could mean that you'll have a lack of instrument definition as a result. The correct angle is determined strictly by personal preference. Some mixers prefer the monitors to be angled directly at their mixing position and others prefer the focal point (the point where the sound from the tweeters converges) anywhere from a foot to about 3 feet behind them to eliminate some of the "hype" of the speakers.

TIP: A great trick for getting excellent left/right imaging is to mount a mirror over each tweeter and adjust the speakers so that you can see your face clearly in both mirrors at the same time when you are in your listening position.

- **Check how the monitors are mounted.** Monitors that are mounted directly on top of a desk without any decoupling are subject to comb-filter effects, especially in the low end. That is, the sound from the monitor causes the desk or console to resonate, causing both the desk and the speaker to interact as certain frequencies either add or subtract. This effect, known as phase cancellation, causes a subtle yet very real blurring of the sound. As a result, it will be just a little harder to hear your low end distinctly, which makes it more difficult to EQ. Phase cancellation can be more or less severe depending on whether the speakers are mounted directly on the desk or metal meter bridge or they are mounted on a piece of decoupling material.

- One of the quickest ways to improve the sound of your monitor system is to **decouple your speakers** from whatever they're sitting on. This can be done with a commercial product, such as Primacoustic's Recoil Stabilizers (see Figure 4.2), or you can make something similar relatively cheaply with some open-cell (closed-cell will work, too) neoprene or even some mouse pads.

Figure 4.2: NS10 with decoupler
Courtesy Prime Acoustics

Decoupling your subwoofers (if you're using them) from the floor can really help too. Although sometimes the coupling with the floor can make your low end feel bigger, it will be a lot clearer and distinct if decoupled. Auralex even offes a product for this called the SubDude-II, although you can probably put together a DIY setup that can work just as well.

Regardless of the brand, model, and type of speakers you use, decoupling is a cheap and easy way to improve your sound right away.

TIP: The best solution is to mount your monitors on stands just directly behind the desk or meter bridge. Not only will this improve the low-frequency decoupling, but it can greatly decrease the unwanted reflections off the desk or console.

- **Check how the monitor parameters are set.** Many monitors are meant to be used in an upright position, yet users frequently will lay them down on their sides. This results in a variety of acoustic anomalies that deteriorate the sound (check your speaker's manual for the manufacture's recommendation). Also, with powered monitors, be sure that the parameter controls of both monitors are set correctly for the application and are the same on each (see Figure 4.3).

Figure 4.3: Monitor speaker parameter controls
© 2024 Bobby Owsinski

- **Check the position of the tweeters.** Most engineers prefer that the tweeters of a two- or three-way speaker system be on the outside, thereby widening the stereo field. Occasionally, tweeters to the inside work, but this usually results in a smearing of the stereo image. Experiment with both, however, because you never know exactly what will work until you try it (see Figure 4.4).

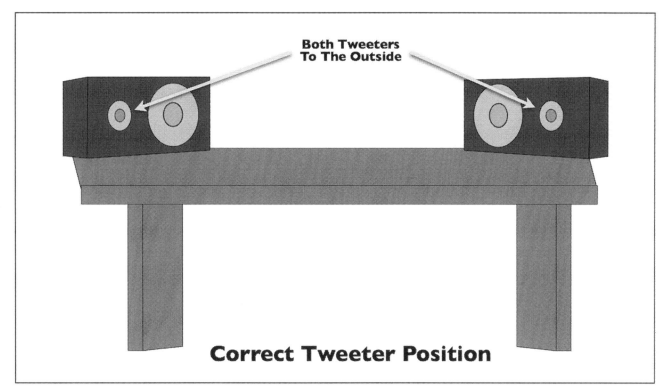

Figure 4.4: Tweeter position
© 2024 Bobby Owsinski

On the Bottom

Mastering engineers take pride in getting the low end of a project right so that it translates well on speaker systems of all sizes. This is one reason why nearfield monitors or even popular soffit-mounted large monitors are inadequate for mastering.

You can only properly tune the low end of a track if you can hear it, so a monitor with a frequency response down to at least 40Hz is essential for accurately working on low frequencies.

To hear that last octave on the bottom, many mastering engineers are now adding subwoofers to their playback systems. A great debate rages as to whether a single subwoofer or stereo subwoofers are required for this purpose. Those that say stereo subs are a must insist that enough directional response occurs at lower frequencies to require a stereo pair.

Stereo subs also a sense of envelopment that better approximates the realism of a live event. Either way, the placement of the subwoofer(s) is of vital importance due to the standing waves of the control room at low frequencies.

Three Steps To Adding A Subwoofer

It's not unusual for musicians and engineers to crave more bottom end for the speakers they're using in their personal studios. As a result, the first thing they think about is adding a subwoofer to their monitor system.

That's all well and good, but there are a few steps you can follow that might make your venture into low-frequency territory a lot easier.

1. Do you really need a subwoofer? Before you make that purchase, make sure a sub is actually necessary. Here are a couple of things to check out first:

- **Are you monitoring at a loud enough level?** This is a trap that people with home studios fall into, as they don't listen loudly enough, even for short periods of time. First of all, if your monitor speakers are too quiet, your ears begin to emphasize the mid-frequencies. This is great for balance but bad for judging the low end of a song. *Your monitors should be at least loud enough that you can adequately hear the low-end of the frequency spectrum.* If you still don't have enough low end, go on to the next point.

- **Do you have an acoustic problem in your room?** Chances are that either your monitors are too close to the wall or they're placed at a point of the room length where standing waves cause some of the low end to cancel out. This is more likely to affect one area of the low-frequency spectrum rather than the entire low end. Just to be safe, *try moving your speakers a foot or so backward and forward to see whether the low end returns*. If not, move on to Step 2.

2. Purchase a subwoofer from the same manufacturer as your main monitors. The easiest way to get a smooth-sounding low end that doesn't cause you more grief that it's worth is to buy a sub to match the monitors that you use most of the time. That means if you're using JBLs, choose a JBL sub that's made specifically for that system; if you're using Genelecs, do the same; KRKs, the same, and so on. This will make a huge difference, especially at the crossover frequency point where the main speakers cross over to the sub. It's usually extremely difficult to get that area to sound natural if you mix brands.

3. Calibrate your sub correctly. Most musicians and engineers that choose to use a sub just randomly dial in the level. You might get lucky and get it right, but it's more than likely that your level will be off, causing a number of unbalanced-sounding mixes until you finally figure it out.

Calibrating Your Subwoofer

1. With the subwoofer bypassed, send pink noise to your main monitors. At your listening position and while listening to one monitor only, use an SPL meter (just about any of them will do to get you in the ballpark, even a phone app) and adjust the level of the monitor until it reads 85dB. The SPL meter should be set to the C Weight and Slow settings. Repeat on the other channel and set that so it also reads 85dB.

2. Turn off the main monitors. Send pink noise just to the subwoofer. Set the level of the SPL meter so it reads 79dB. Although it may seem like it will be lower in level, 79dB works because there are fewer low frequencies bands than high (three for the low and eight for the high), so this adjustment takes that into account. You might have to tweak the level up or down a dB, but this will get you into the ballpark.

3. If your subwoofer has a polarity switch, try both positions and see which one has the most low end or sounds the smoothest in the crossover area. That's the one to select.

Following these steps will make integrating a subwoofer into your system as easy as possible, if you decide you need one.

Placing The Subwoofer

Here's a method that will get you in the ballpark, although you'll have to do a bit of experimenting. Keep in mind that this method is for single subwoofer use.

1. Place the subwoofer in the engineer's listening position.

2. Send pink noise only into the subwoofer at the desired reference level (85dB SPL should do it, but the level isn't critical).

3. Walk around the room near your main monitor speakers until you find the spot where the bass is the loudest. That's where you should place the sub. For more low-end level, move it toward the back wall or corner, but be careful because this could provide a peak at only one frequency. We're looking for the smoothest response possible (which may not be possible without the aid of a qualified acoustic consultant).

Amplifiers

While most recording-style monitor speakers tend to be self-powered, many preferred speakers in professional mastering environments are passive. As a result, they still require an outboard amplifier—

and often a large one. It's not uncommon to see amplifiers of well over 1,000 watts per channel in a mastering situation.

This is not for level (since most mastering engineers don't listen all that loudly), but more for headroom, so that the peaks of the music induce without a hint of distortion. Since many speakers used in mastering are rather inefficient as well, this extra amount of power can compensate for the difference.

Although many power amps that are standard in professional recording, such as Manley, Bryston, and Crown, are frequently used, audiophile units such as Cello, Threshold, Krell, and Chevin are also common.

> *When I started Gateway, I got another pair of Duntech Sovereigns and a new pair of Cello Performance Mark II amplifiers this time. These are the amps that will put out like 6,000-watt peaks. One never listens that loudly, but when you listen, it sounds as though there's an unlimited source of power attached to the speakers. You're never straining the amp, ever.*
>
> —Bob Ludwig

LISTENING TECHNIQUES FOR MASTERING

Regardless of what kind of monitors or room you have to work with, there are some proven techniques that will yield reasonable results even under the worst conditions. These all depend upon your ears, which are still the primary ingredient in mastering, and not the gear.

- **Start by listening to some high-resolution music that you love.** Aim to listen to the highest quality audio possible, at least CD quality. A FLAC, high-res WAV, or AIFF file is even better. Listen to a favorite recording or two that you know really well and understand how it sounds on your system so that you can establish a reference point. This will keep you from over-EQing or compressing too much. If you do nothing else, this one trick will help you more than anything else.

- **Establish a different listening level.** You need one level where you can easily hear how the lower-frequency instruments (especially bass and drums) sit with each other. Although the SPL level may vary from mastering engineer to mastering engineer, television uses 82dB SPL and film uses 85dB SPL and never vary.

- **Stick to this listening level only.** Mark it down on your volume control, make a note where the level is in the software, and do whatever you have to do to make this level repeatable. The level is somewhat arbitrary in that it depends on your monitors and your environment, but the idea is that you want a level that's loud enough for you to gauge the low end and hear the tonal balance. If you listen at varying levels, your reference point will be thrown off, and you'll never be sure exactly what you're listening to, which is why you keep it to one level only.

- **Use two sets of speakers: one large set and one small.** The only way you can ever be sure of how things really sound is if you have two different speaker sets to reference against, especially if you don't have large super-high-resolution reference monitor speakers. Even if the largest speaker system that you can afford is a two-way bookshelf speaker with a 6-inch woofer, you should have an even smaller set to reference against. Although not the best, even a pair of computer speakers will do as long as you can feed them from the same source as your larger monitors. Mastering pros usually use a huge set of monitors with double 15-inch woofers plus a subwoofer, and an average two-way bookshelf speaker or something even smaller. Even if you have more than two sets of monitors available, limit your listening choices during mastering so you don't confuse yourself and end up chasing your tail.

- **When mastering your own mix, use a different set of speakers from the ones you mixed on.** It doesn't matter whether you mixed in your bedroom or in a million-dollar SSL room with George Augspurger acoustics, you're at a huge disadvantage if you try to master using the same monitors that you mixed on. Why? Because all monitors have flaws, and if you use the same monitors for mastering, you're either overlooking a frequency problem or just making it worse. This is important because if you use the same monitors, you'll only be compounding any frequency response problems that the speakers might have in the first place.

TIP: Some projects are better off being remixing rather than trying to fix during mastering. Don't be afraid to send the project back (or to redo it yourself, if you're the mixer) rather than spending a lot of time making it sound different, not better. Believe it or not, pro mastering engineers make this suggestion all the time, even to some of the top mixers.

Monitors Versus Headphones

Sometimes it's just not possible to listen to your monitors when you're working on music at home. When it's late at night and your kids, significant other, or neighbors are in the next room, separated only by paper-thin walls, you have no choice but to try to listen on headphones.

Mastering (or mixing, for that matter) on standard headphones does have four significant downsides, though.

- **Your ears get tired.** You can't wear headphones for extended periods (8, 10, 12 hours) before your head and ears get tired from the extra weight.

- **It's easy to get ear fatigue.** You have a tendency to turn them up, which can lead to some quick ear fatigue, again limiting your ability to listen for long periods.

- **The headphones can give you a false sense of what the song really sounds like.** Because most of the more expensive professional headphones really sound great, you get a false sense of what you're listening to (especially on the low end), and it causes you not to work as hard at getting the frequency balance right.

- **If you master something on headphones, it might not translate to speakers.** If you master something only on headphones, it might not work when played back on normal monitors.

Room Simulation

When the last edition of this book came out I would have recommended that you still need to do most of your work on speakers to be sure that it will translate to a playback medium of any type. While that's still the preference, today room simulation and correction software give you a real chance to create something useful using just headphones.

Room correction systmes like Sonarworks and IK Multimedia's ARC 4 can not only improve the acoustics of your room, but your headphones as well. Room simulation software from Waves, Slate Digital VSX, Acoustica and others now make mixing and mastering possible without monitors.

Is it better than using a great set of speakers in a great playback environment? No, but it will get you closer to a usable final result than headphones ever could before.

Figure 4.5: Steven Slate VSX
Courtesy of Slate Digital

TIP: When using room correction software, the more samples of the room that you take, the better the result.

TAKEAWAYS

- Your acoustic environment plays a crucial role in the performance of your monitor speakers.
- Mastering engineers often spare no expense when selecting monitor speakers.
- Proper monitor speaker placement is a simple and cost-effective way to significantly improve playback performance.
- It's critical to calibrate a subwoofer to truest playback performance.
- With room correction software, the more samples you take, the better the result.
- It's possible to mix on headphones using control room simulation software.
- Redoing a mix can be faster and provide a better result than trying to fix some issues during mastering.

5

MASTERING TOOLS

All tools created for mastering, regardless of whether they're analog or digital, have two major features in common: extremely high sonic quality and repeatability. The sonic quality is a must in that any device in either the monitor or the processing chain should have the least possible effect on the signal.

Repeatability is important (although less so now than in the days of vinyl) as the exact settings must be repeated in the event that a project is redone, as in the case when additional masters or changes are called for weeks later.

While this feature isn't much of a problem in the digital domain because the settings can be memorized, many analog mastering devices are still used in pro facilities. As a result, these hardware devices require special mastering versions that have 1dB or less indented increment selections on the controls, which can seriously add to the cost of the device.

That said, the mastering that you'll most likely do will probably all be "in the box," and there are a variety of more affordable options to choose from. In this chapter, we'll look at each tool as well as their ideal placement in the audio signal path.

THE COMPRESSOR

In mastering, the compressor is the primary way of raising the relative level of the program and giving the master both punch and strength. Relative level is how loud we perceive the volume rather than the absolute level that's on the meter.

Compressor Overview

A compressor is a dynamic level control that uses the input signal to determine the output level. The *Ratio* parameter controls the amount that the output level from the compressor will increase compared to the input level (see Figure 5.1).

For instance, with a compression ratio of 4:1 (four to one), for every 4dB of level that goes into the compressor, only 1dB will come out once the signal reaches the threshold level (the level at which the compression begins). If a gain ratio is set at 8:1, then for every 8dB that goes into the unit, only 1dB will come out its output.

Some compressors have a fixed ratio, but the parameter is variable on most units from 1:1 (no compression) to as much as 100:1 or more (which makes it a limiter, a process that we'll look at later in this chapter). A *Threshold* control sets the input-level point where the compression will kick in. Under that point, no compression occurs.

Figure 5.1: Compressor Attack and Release controls
© 2024 Bobby Owsinski

Most compressors have *Attack* and *Release* parameters. These controls determine how fast or slow the compressor reacts to the beginning (attack) and end (release) of the signal envelope. Many compressors have an *Auto* mode that automatically sets the attack and release according to the dynamics of the signal.

Although *Auto* mode works relatively well, it doesn't allow for the precise settings required by certain source material. Some compressors (such as the revered Teletronix LA-2A or dbx 160) have a fixed attack and release, which helps give the compressor a distinctive sound.

When a compressor actually compresses the signal, the level is decreased, so there's another control called *Make-Up Gain* or *Output* that allows the signal to be boosted back up to its original level or beyond.

Using The Compressor In Mastering

In mastering, the compression ratio of the mastering compressor is usually set very low, from between 1.5:1 to 3:1, in order to keep the compression fairly gentle-sounding. The higher the ratio, the more likely you'll hear the compressor work, which can cause the program to sound unnatural. That might be a good thing sometimes in recording and mixing where a certain color is desired, but mastering is trying to keep the sound of the original program as intact as possible.

The keys to getting the most out of a compressor in mastering are the *Attack* and *Release* controls, which have a tremendous overall effect on a mix and therefore are important to understand. Generally speaking, transient response and percussive sounds are affected by the *Attack* control setting. *Release* is the time it takes for the gain to return to normal after compression occurs.

In a typical pop-style mix, a fast *Attack* setting will react to the drums and percussion and reduce the overall gain on each beat. If the *Release* is set very fast, then the gain will return to normal quickly, but can have an audible effect by reducing some of the overall program level and attack of the drums in the mix.

As the *Release* is set faster, the gain changes more quickly, which might cause the drums to "pump," meaning that the level of the mix will increase and then decrease noticeably. Each time the dominant instrument starts or stops, it "pumps" the level of the mix up and down.

Mastering compressors that work best on a wide range of full program material generally have very smooth release curves and slow release times to minimize this pumping effect. Chapter 7, "Mastering Techniques," will further detail the effects that *Attack* and *Release* settings can have on the program.

Widely used mastering compressor plugins include the Shadow Hills Mastering Compressor, PSP MasterComp, and the various versions of the Fairchild 670.

Hardware compressors include the Manley SLAM Master, the Shadow Hills Mastering Compressor, and the Maselec MLA-2.

Multiband Compression

Multiband compression splits the input audio signal into several frequency bands, each with its own dedicated compressor. The main advantage of a multiband compressor is that a loud event in one frequency band won't affect the gain reduction in the other bands.

That means that something like a loud kick drum will cause the low frequencies to be compressed, but the mid and high frequencies are not affected. This allows you to get a more controlled, hotter signal with far less compression than with a typical single-band compressor.

The multiband compressor is unique in that it can shape the timbre of a mix in ways that an EQ just can't. By raising or lowering the *Level* control of each band and then adjusting the *Crossover* controls to adjust the bandwidth, the tonal quality of the mix changes similarly to using an EQ, but with the compression applied only to the selected band.

For instance, if you wanted to tighten up the low end, you'd increase the level of the low-frequency band while adjusting the low/mid frequency Crossover control to pinpoint the exact frequencies you want to affect.

Frequently used multiband software compressors include the Waves C6 (see Figure 5.2), the Universal Audio UAD Precision Multiband, the Solid State Logic G3 MultiBus compressor, or the multiband compressor in iZotope Ozone. Hardware multiband compressors include the Maselec MLA-3 and the Tube-Tech SMC 2BM.

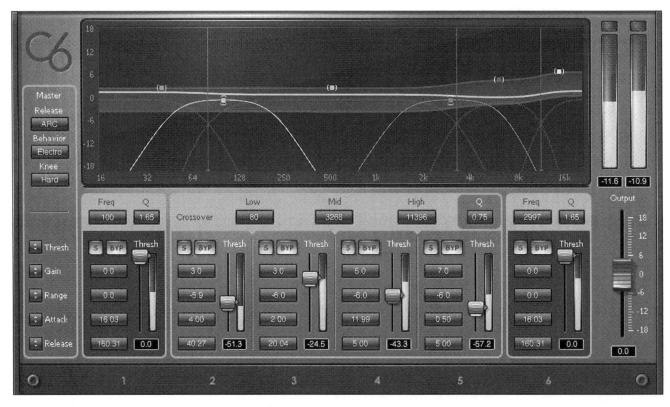

Figure 5.2 : Waves C-6 Multiband Compressor plugin
Courtesy Waves

The Dynamic Equalizer

It's worth noting that multiband compressors are being used less frequently thanks to the introduction of the dynamic EQ. Just like a typical equalizer, each band can be boosted or cut in varying degrees depending on how you want to shape the incoming signal. However, a dynamic EQ automates the intensity of those boosts and cuts depending on the characteristics of the incoming signal, including frequency content distribution and loudness.

The dynamic EQ has all of the parameters of the typical multiband compressor, but it also includes a Q (bandwidth) parameter that allows the mastering engineer to determine how many frequencies each band will affect. Most dynamic EQs also have more bands available than the typical multiband compressor as well.

Frequently used multiband software compressors include the FabFilter Pro-Q3 (see Figure 5.3), the Hofa IQ-Series EQ, and the Sonnox Oxford Dynamic EQ.

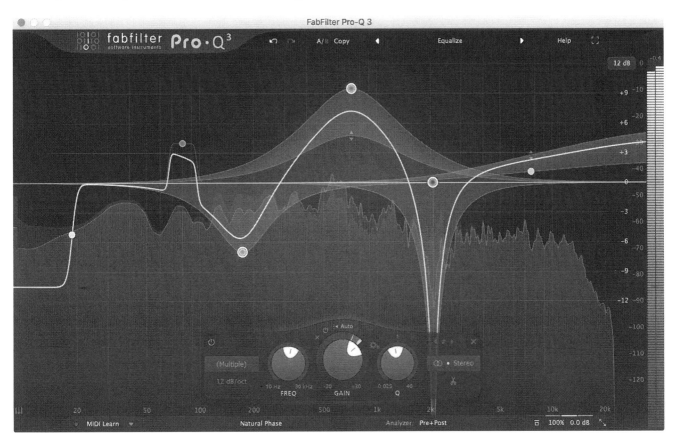

Figure 5.3 : The FabFilter Pro-Q 3 dynamic EQ
Courtesy FabFilter

THE LIMITER

One of the most essential tools for a mastering engineer is the limiter, since it's the primary way that high levels are achieved without digital overs.

Limiter Overview

A limiter is a compressor with a very high compression ratio and a very fast attack time, allowing it to capture the quick peaks of an audio signal. Any time the compression ratio is set to 10:1 or more, the result is considered limiting. While it's true that a compressor can be adjusted to work as a limiter, in mastering the limiter is usually a dedicated unit created specifically for the task.

A mastering limiter can be thought of as a brick wall for level, allowing the signal to get only to a certain point and little more. Think of it like a governor that's sometimes used on trucks to make sure that they don't exceed the speed limit. After you hit 65 mph (or whatever the speed limit in your state is), no matter how much more you depress the gas pedal, you won't go any faster.

It's the same with a limiter. Once you hit the predetermined level, no matter how much more signal is fed into the limiter, the level virtually stays the same.

Using The Limiter In Mastering

To understand how a limiter works in mastering, you have to understand the composition of a typical music program first. In general, the highest peak of the source program (the song in this case) determines the maximum level possible in a digital signal.

Because many of these upper peaks have a very short duration, they can usually be reduced in level by several dB with minimal audible side effects. By controlling these peaks, the entire level of the program can then be raised several dB, resulting in a higher average signal level.

Most digital limiters used in mastering are known as *brick-wall limiters*. This means that no matter what happens, the signal will not exceed a certain predetermined level, and there will be no digital overs after its *Ceiling* level has been set. A brick-wall limiter is usually set anywhere from –0.1dB to –1.0dB, and once set, the level will never go beyond that point.

Many popular mastering limiter plugins are commonly used, from the Waves L1 and L2, to the Universal Audio Precision Limiter (see Figure 5.3), to the T-RackS Brickwall Limiter, to Massey L2007 and many others.

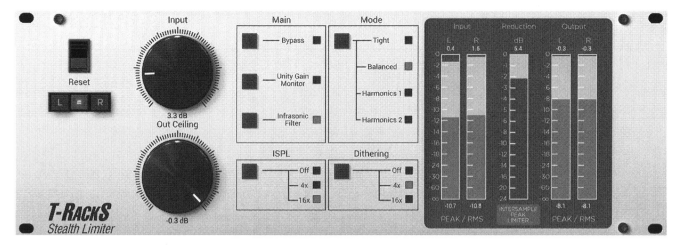

Figure 5.4: T-RackS Stealth Limiter
Courtesy IK-Multimedia

By properly setting a digital limiter, the mastering engineer can increase the apparent level by several dB, simply because the peaks in the program are now controlled.

TIP: Adding too much limiting can make a mix sound lifeless and even fatiguing to listen to.

Multiband Limiter

Multiband limiters work similarly to multiband compressors, splitting the audio spectrum into separate frequency bands that can be individually limited. This can provide more level without sounding as compressed as when a single-band limiter is used. The operations are identical to that of a multiband compressor, except that the compression ratio is higher, making it a limiter instead of a compressor.

Commonly used mastering multiband limiter plugins are the Waves L3, iZotope Ozone Multiband Limiter, and McDSP MC2000 and ML4000, among others.

THE EQUALIZER

One of the most important duties of the mastering engineer is correcting the frequency balance of a project if it's required. Of course this is done with an equalizer, but the type used and the way it's applied are generally far different than when used during recording or mixing.

Most precision work in mastering is done using a parametric equalizer, which allows the engineer to select a frequency, adjust the amount of boost and cut, and control the bandwidth around the selected frequency that will be affected (known as *Q*).

Using the EQ in Mastering

Where in recording you might boost or cut large amounts of EQ anywhere from 3 to 15dB at a certain frequency, mastering is almost always done in very small increments, usually in tenths of a dB to 2 or 3dB at the very most. What you might see are a lot of small shots of EQ along the audio frequency band, but again in very small amounts.

For example, these might look like +1dB at 30Hz, +0.5dB at 60Hz, +0.2dB at 120Hz, −0.5dB at 800Hz, −0.7dB at 2500Hz, +0.6dB at 8kHz, and +1dB at 12kHz. Notice that there's a little added to a lot of places across the frequency spectrum.

Another technique that's used frequently is known as *feathering*. This means that instead of applying a large amount of EQ at a single frequency, small amounts are added at the frequencies adjoining the one being focused on. An example of this would be instead of adding +2dB at 100Hz, you would add +1.5dB at 100Hz and +0.5dB at 80 and 120Hz (Figure 5.5). This generally results in a smoother sound by not stressing any one area of the equalizer.

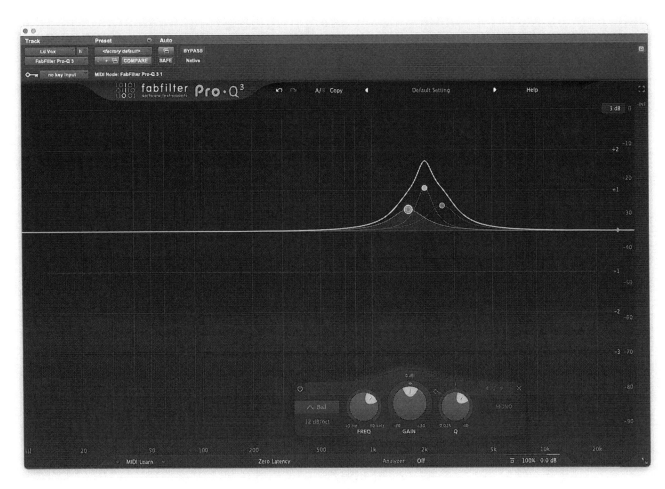

Figure 5.5: EQ feathering
© 2024 Bobby Owsinski (Source: FabFilter)

In mastering, if large amounts of EQ are required it's usually an indication that there's something wrong with the mix. Top mastering engineers often send a mix back to be redone, as a corrected mix will typically sound better than one requiring major EQ surgery. With mastering equalization, less is definitely more.

Hardware mastering equalizers differ from their recording counterparts in that they usually feature stepped rather than continuously variable controls in order to be able to repeat the settings, with the steps being in increments as small as 0.5dB, although 1dB increments are the most common. Examples are the Manley Massive Passive (see the mastering version in Figure 5.6) and the Avalon AD2077.

Figure 5.6: The Manley Massive Passive mastering version
Courtesy of Manley Labs

Most equalizer plugins are inherently capable of 1dB steps, so the decision often comes down to the audio quality each one imparts to the program. EQs that impart their own "color" to the audio are usually passed over in favor of more transparent versions that don't change the sonic quality of the program very much. Examples are the Massenburg DesignWorks MDWEQ5, the Sonnox Oxford EQ, the Brainworx bx_digital V2, the PSP Neon HR, and many more.

TIP: Many equalizers offer a high-resolution mode, called Linear Phase, which increases the smoothness of the EQ. If available, this setting reduces the sonic coloration of the equalizer.

THE DE-ESSER

Sibilance is a short burst of high-frequency energy that over-emphasizes the 'S' sounds in vocals. It usually results from a combination of the vocalist's mic technique, the type of mic used, and heavy compression on the vocal track or mix buss. Sibilance is undesirable, so a special type of compressor called a *de-esser* is used to suppress it.

Most de-essers have two main controls, *Threshold* and *Frequency*, which are used to compress only a very narrow band of frequencies, typically between 3kHz and 10kHz, to eliminate sibilance (see Figure 5.7).

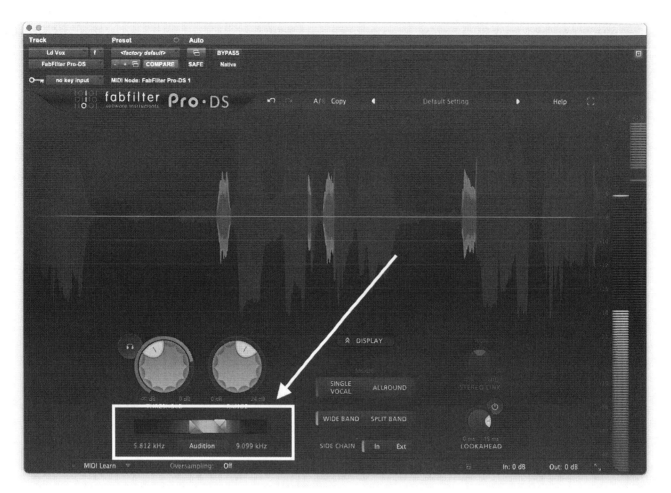

Figure 5.7: Sibilance band between 3k and 10kHz
© 2024 Bobby Owsinski (Source: FabFilter)

Modern software de-essers are much more sophisticated than analog hardware de-essers, but the bulk of the setup still revolves around those two parameters. A common feature in modern de-essers is a *Listen* button lets you solo only the frequencies being compressed, helping you pinpoint the exact band that's causing the issues (see Figure 5.8).

Figure 5.8: A de-esser with the Listen function
© 2024 Bobby Owsinski (Source: Avid)

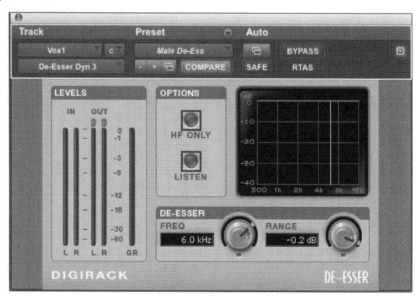

While de-essers are typically used for sibilant vocals, they can also be useful for controlling excessive high frequencies from other instruments as well. Cymbals, guitars, and even the snare drum can occasionally benefit from this unique tool.

Typical de-esser plugins include the Massey De:Esser, McDSP DE555, and Waves DeEsser, among others.

TIP: Sometimes a dynamic EQ can be used as a compressor since it can de-ess in multiple bands at the same time.

SATURATION

In the analog domain, saturation occurs when the electronic components inside an audio device are overloaded. When an electronic component like a transistor, transformer, vacuum tube or magnetic tape is not able to handle the level of the incoming audio signal, then the output of the device results in a subtle compression and distortion. In some cases this can add a pleasant effect to the audio, but if pushed too far can become distinctly non-musical.

Types Of Saturation

The type of distortion that most overloaded audio devices add is called harmonic distortion, which adds small spikes of amplitude to multiples of the incoming frequency.

For instance, even-order harmonics, which we generally find warm and pleasant, applied to a 100Hz sine wave would produce harmonic spikes at 200Hz, 400Hz, and 600Hz. Gear built around vacuum tubes have even-order distortion, which is why we like the sound of tube gear so much.

On the other hand, odd-order harmonics of a 100Hz sine wave would produce amplitude spikes at 300Hz, 500Hz, 700Hz, and so on. Odd-order harmonics have a much more aggressive, bolder sound. Gear based around transistor or "solid state" components tend to have more odd-order distortion.

It's interesting that magnetic tape has a combination of even and odd order harmonics, but it's the gentle rolloff of the high frequencies that produces the warmer sound that we like. Don't forget that the electronics tied to magnetic tape recording (almost always solid state) impart their own sound to the saturation as well.

Saturation also introduces a compression component. When a device just goes into saturation the compression is around a 2:1 ratio, but if you drive the component harder, the ratio increases to as much as 4:1.

Soft Clipping

We've all heard the type of digital overload where a nice clean sound suddenly breaks into a crackle of unpleasantness. This happens in an instant and is the major reason why everyone wanted to stay away from digital overloads in the early days of digital audio.

As digital gear design progressed, mastering engineers noticed that some devices clipped more gently, leading to the term 'soft clipping." Soft clipping is very analog-like in that it very gradually eases into saturation so the harsh digital clipping of the past is diminished. Most digital processors today have some form of soft clipping built in, although not all sound particularly pleasant to the ear.

Saturation Uses

Some mastering engineers are staunchly against adding anything that adds distortion to the audio signal and won't be caught dead using a saturator under any circumstances. Others feel that if it helps them achieve the result that the client is looking for, then adding a little bit of intentional distortion is fair game in getting the job done.

That said, saturation (especially soft clipping) is often used when the mastering engineer wants to increase the level slightly without adjusting the compressor or limiter.

Sometimes adding even-order harmonics can enhance the low-end of a mix, while odd-order harmonics can help a bass-heavy mix cut through on small speakers. Finally, tape saturation is often used for adding some additional "glue" to a mix while increasing the perceived loudness.

TIP: Many plugins use upsampling, which raises the sample rate to allow more precise digital processing, resulting in a cleaner sound.

Saturation Plugins

Like all but the latest plugins, saturation plugins are modeled on analog gear, and attempt to emulate the same harmonics of the analog device when overloaded. In some cases, the saturation will start out with mostly even harmonics but include more odd harmonics as it's driven harder.

Just about every plugin developer has a saturation plugin offering, but popular ones include Soundtoys Decapitator, FabFilter Saturn2, Schwabe Digital Gold Clip, and UAD Studio A800.

CONVERTORS

Digital audio requires a device to convert the analog signal into a digital stream of 1s and 0s, and then after the digital audio has been recorded and processed, return it back into an analog signal that we can listen to. These devices are called analog-to-digital (A/D or ADC for short) and digital-to-analog (D/A) convertors.

Mastering studios are especially concerned with the quality of the D/A convertor (also known as a *DAC*), since the highest quality signal path to the monitors is a priority. As a result, the built-in convertors of most commonly used audio interfaces are insufficient, so most facilities opt for stand-alone outboard convertors.

A key criterion for a mastering DAC is the ability to operate at a wide range of sample rates, from 44.1kHz to 192kHz, which most devices on the market currently support. In the future, 384kHz may also be a required option.

Because each brand has a slightly different sound (just like most other pieces of gear), major mastering facilities may have numerous versions of each type available for a particular type of music. Popular converters include Prism Sound, Lavry Engineering (see Figure 5.23), Apogee, and Benchmark Media, among others.

Figure 5.9: Lavry 122-96 analog-to-digital converter
© 2024 Bobby Owsinski

Unless you're mastering from an analog source like tape, or using analog processing outside of the DAW with an outboard analog device, external analog-to-digital convertors are not necessary for mastering.

CONSOLES/MONITOR CONTROL

While mastering consoles (sometimes referred to as *transfer* consoles) were once the centerpiece of the mastering studio, they are now used primarily to switch between different input and speaker sources and adjust the monitor level. That said, once again the emphasis is on a very high-quality signal path that degrades the signal as little as possible.

In the analog days when vinyl records were dominant, a mastering console was a sophisticated device often featuring two sets of equalizers and, in many cases, built-in compression. As mastering entered the digital age, mastering consoles were among the first pieces of gear to become fully digital, with all EQ, compression, and limiting handled on-board.

Since the vast majority of processing is done in the DAW these days, all that's required is the monitor section to control the volume level of the monitors, switch between different monitors, and control input sources.

As with the D/A convertor, a key criterion for a mastering monitor controller is the ability to support a wide range of sample rates, from 44.1kHz to 192kHz, which most devices on the market currently do. In the future, 384kHz may also be a required option.

Due to the unique nature and relatively small size of the mastering market, not many companies currently manufacture dedicated mastering consoles/monitor controllers. Among the manufacturers are Crookwood, Maselec, Dangerous Music, and Manley Labs.

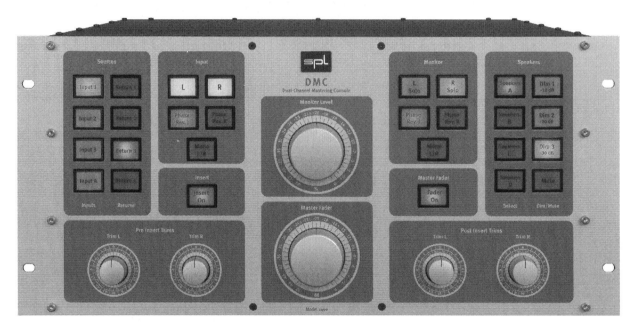

Figure 5.10: The SPL DMC
Courtesy of SPL

THE DIGITAL AUDIO WORKSTATION

The digital audio workstation (DAW) has become the heart and soul of the mastering studio, enabling engineers to perform tasks like editing and sequencing with much greater ease than was once thought possible. Plus, the DAW enables new processing tasks that were unimaginable a decade ago.

Mastering DAWs

Although any DAW can be used for mastering in a pinch, a few manufacturers have become favorites among mastering engineers, primarily because of dedicated mastering features that are included. These features are mostly CD-specific and include DDP export, dither, and PQ code insertion and editing, all of which will be covered in Chapter 12, *Mastering for CD*.

DAWs that provide these functions include Steinberg WaveLab, Sonoris DDP Creator, DSP-Quattro, and SADiE, HOFA's specialized mastering DAW and plugin set CD-Burn.DDP.Master PRO, among others.

There are also a number of dedicated software mastering suites available that provide precision processing for mastering, including iZotope Ozone and Waves, among others. Standalone mastering apps on the computer desktop include Roxio Toast, CD Architect, and T-RackS.

It should be noted that some mastering engineers may use two separate DAW setups—one for playback, and another for processing. Usually a Pro Tools DAW is used for playback and the DAW of choice for the mastering engineer for processing.

OTHER TOOLS

There are also other tools sometimes used in the mastering environment that don't fall under one of the previous categories.

Stereo Enhancement

Many times the mix seems like it's either too wide or too narrow, so a stereo enhancer is called for to adjust the width of the stereo field. This is more of a fix-it tool than something that a mastering engineer would use every day, but it does come in handy during those rare times when the stereo field needs adjustment.

Examples include the PSP StereoEnhancer and StereoController, iZotope Ozone, Waves Center, and S1 Stereo Imager.

M-S Processing

Some processors feature an M-S mode, which uses a mid-side matrix that allows you to select where the process works within the stereo image by manipulating the in and out-of-phase signals.

The Mid assigns the processing to the center of the stereo image as usual, while the Side makes it seem as if the processing is occurring at the outside edges of the stereo sound field. Most of the time you'll find that the Mid mode works best, but occasionally Side mode can be more effective, such as brightening a track without affecting a lead vocal.

M-S processing can be found on many T-RackS processors as well as Brainworx Control V2, among others.

Mono

As previously stated earlier in this chapter, it's always a good idea to check your work in mono. If you have an outboard monitor controller, it likely has a mono switch, but if that's not something yet in your gear arsenal, then you need to switch to mono in software.

If the processors in your mastering chain don't provide this ability, a great little utility is Brainworx bx_solo (and it's free), which can also be used for stereo width adjustment.

TAKEAWAYS

- The compression ratio of the mastering compressor is typically set between 1.5:1 to 3:1 in order to maintain a fairly gentle compression sound.

- The higher the ratio, the more likely you'll hear the compressor work, which can make the program sound unnatural.

- Most compressors have *Attack* and *Release* parameters, that control how fast or slow the compressor reacts to the beginning (attack) and end (release) of the signal envelope.

- When the *Release* is set faster, the gain changes more quickly, which might cause the level of the mix to fluctuate noticeably.

- A dynamic EQ shares many of the parameters of the typical multiband compressor, but also includes a Q (bandwidth) control that allows the mastering engineer to adjust the frequency range each band affects.

- A limiter is a type of compressor with a high compression ratio and fast attack specifically designed to catch quick peaks in an audio signal. When the compression ratio is set to 10:1 or more, the result is considered limiting.

- Most digital limiters used in mastering are called *brick-wall* limiters, meaning the signal cannot exceed a certain predetermined level, and there will be no digital overs after its *Ceiling* level has been set.

- A brick-wall limiter's *Ceiling* control is usually set anywhere from −0.1dB to −1.0dB, and once set, the level will never go beyond that point.

- Mastering EQ is almost always done in very small increments, usually in tenths of a dB, to 2 or 3dB at the very most.

- A de-esser is used to suppress sibilance by compressing a very narrow band of frequencies typically between 3kHz and 10kHz.

- Magnetic tape saturation emulators generate both even and odd-order harmonics, but it's the gentle high-frequency rolloff that creates the warmer sound we appreciate.

- Saturation includes a built-in compression component, starting at a 2:1 ratio and increasing to as much as 4:1 when driven harder.

- Soft clipping mimics analog behavior by gradually easing into saturation, reducing the harshness of traditional digital clipping.
- Saturation, particularly soft clipping, is often used by mastering engineers to slightly increase the level without adjusting the compressor or limiter.

METERING TOOLS

Metering is extremely important in mastering, much more so than in mixing, especially when you're aiming for hot levels. There are more metering tools available to the mastering engineer than the simple metering that we're used to using during recording, because the mastering process requires a lot more visual input to tell you the things you need to know.

While you don't need all of the following meters to do a proper mastering job, they all do serve a purpose. You'll find that they'll all contribute to making a great master. Let's look at some of the meters that are frequently used.

THE PEAK METER

The peak meter was created by the BBC when they realized that the common VU meter (see Figure 6.1) wasn't showing the engineer the exact program signal. This is especially important in broadcasting, where over-modulating the signal can result in a fine from the Federal Communications Commission in the United States.

The standard analog VU meter (which stands for Volume Units), which was common on all professional audio gear until the late '90s, *only shows the average level of a signal* and has a very slow response time. As a result, you'd have to guess at the signal peaks, since they were too fast for the meter to accurately read them.

A good example of this effect takes place during recording of a high-pitched percussion instrument such as a triangle or a tambourine, where the signal is almost all peaks. The experienced engineer would record the instrument at a barely visible −20dB on the VU meter to keep the recording from distorting.

Couple the slow response of the VU meter with the fact that it's an analog mechanical device that could easily be knocked out of calibration, and you can see the need for a new metering system.

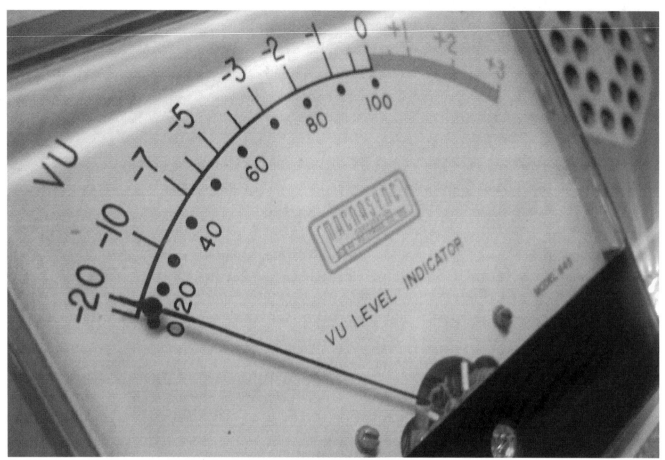

Figure 6.1: A typical VU meter

The peak meter, on the other hand, has an extremely fast response, which is almost fast enough to catch most peaks (more on this in a bit), and could be simulated on a digital display instead of using an actual meter (a hardware peak meter used to be very expensive before the digital age - see Figure 6.2).

The peak meter also became a necessity for digital recording because any signal beyond 0dB could cause anything from a harsh edginess to a very nasty-sounding distortion. As a result, all peak meters have a red *Over* indicator that lets you know you've exceeded the zone of audibly clean level.

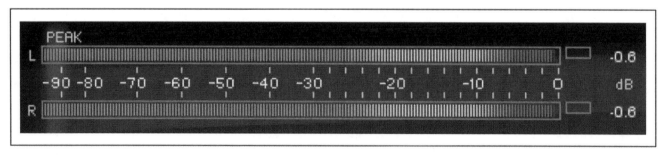

Figure 6.2: A typical digital peak meter
© 2024 Bobby Owsinski (Source: IK Multimedia)

Inter-Sample Distortion

There's also a frequently overlooked phenomenon called inter-sample distortion, where the signal peaks exceed 0dB between the samples of really hot signals and aren't indicated by the Over indicator.

This can cause trouble when the song or program is played back later and the digital-to-analog convertor (D/A) of a device is overloaded even though the Overload indicator never lights (which is why some mastered programs can sound harsh).

Peak meters can also be either somewhat accurate or very accurate, depending upon the resolution of the meter. A peak meter used in mastering would normally be calibrated in tenths of a dB, while some inexpensive implementations might have a resolution of 1dB or more (see Figure 6.2).

The fact is that many standard peak meters can't accurately distinguish between 0dBFS (FS = Full Scale) and an over, which can mean that an overload is occurring without you knowing. That's why mastering engineers rely on super-accurate peak metering that counts the number of samples in a row that have hit 0dB.

This number can usually be set in a preference window from three to six samples. Three overload samples equals distortion that will last for only 33 microseconds at 44.1kHz, so it will probably be inaudible, and even six samples is difficult to hear. That said, this type of peak meter is much preferred during mastering.

THE RMS METER

Even when your peak meter is tickling 0dB, an RMS meter will settle at a point much lower since it's measuring the signal differently. RMS stands for the "root mean square" measurement of the voltage of the electronic signal, which roughly means the average level.

We don't use RMS meters much these days because a peak meter is much more precise, but in the pre-digital days, that's all that was available (since a VU meter is an RMS meter).

Today there are digital versions of the RMS meter (see Figure 6.3). In an app like Blue Cat's DP Meter Pro, the RMS meter combines both left and right channels into a single display that measures the power of a signal.

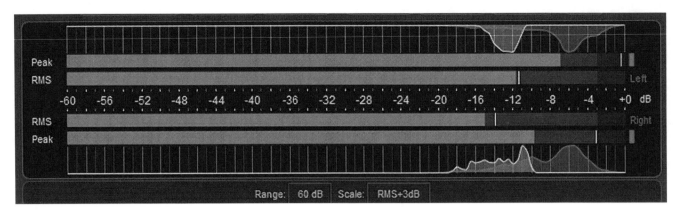

Figure 6.3: Blue Cat DP Meter Pro digital RMS meter
Courtesy Blue Cat Audio

I'd be lying if I said I didn't look at meters, but for me, VU meters are the most important. As someone who started with vinyl, I pay attention to VU meters because I think they're a better way to measure perceived loudness.

—Pete Lyman

Another disadvantage of an RMS meter is its flat frequency response, which can give you a false sense of the level if the song has a lot of low end. That's why it's best to read the RMS meter alongside other meters to get a better sense of both level and loudness (they're different, as you'll soon learn).

TIP: The RMS meter is great for indicating if two (or more) songs are at roughly the same level. However, don't rely on it alone—always use your ears to make the final judgment.

K-System Metering

Mastering engineer Bob Katz has developed a metering system with simultaneous peak and RMS displays that is supported by many metering packages (see Figure 6.4). The K-System features three different meter scales, with the zero point set at -20dB, -14dB, or -12dB, depending on the type of program being worked on. These three scales, K-20, K-14, and K-12, are designed to help engineers maintain enough headroom to avoid clipping.

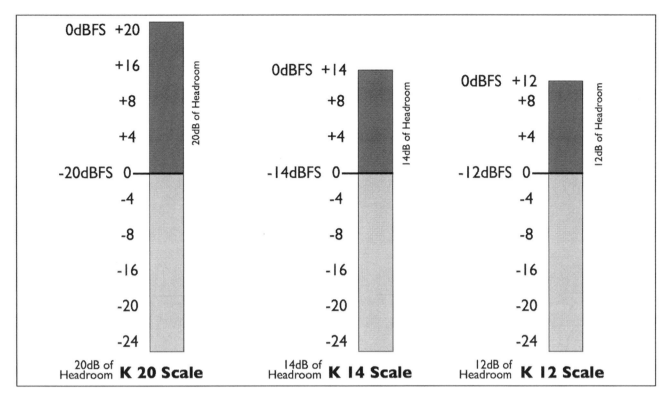

Figure 6.4: A K-System meter
© 2024 Bobby Owsinski

The K-20 shows 20dB of headroom above 0dB and is intended for film and classic mixes. K-14 shows 14dB headroom and is intended for music mixing and mastering. K-12 shows 12dB headroom and is intended for television and radio.

THE PHASE SCOPE

The phase "scope" gets its name from the fact that in the early days of recording, the phase between the left and right channels of a program was checked by using an old-fashioned oscilloscope (see Figure 6.5), which was just called a "scope" for short.

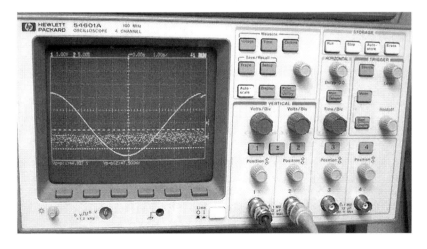

Figure 6.5: A classic oscilloscope
© 2024 Bobby Owsinski

Phase is extremely critical in a stereo signal, because if the left and right channels are not in phase, not only will the program sound odd, but instruments panned to the center (like lead vocals and solos) may disappear if the stereo signal is collapsed into mono.

While you may think that mono isn't used much these days, you'd be surprised to learn that it's actually quite common. If your song is ever played on AM radio, it's in mono on 99 percent of the stations.

On FM radio, if a station is far enough away from where you're listening, the stereo signal may collapse into mono because the signal strength is weak. On television, it's not uncommon for the stereo mix of the show to be automatically converted to mono on some networks.

Sometimes the settings in the Apple Music player are switched to mono, or a stereo song is ripped in mono, meaning it will play back in mono on your device. Mono is everywhere, so it's a good thing to pay attention to the phase of your program.

Using The Phase Scope

The phase scope isn't very good for measuring absolute levels (that's why you have the other meters), but it does provide a wealth of information about stereo source positioning, and the relative phase and level between the two channels.

The signals are displayed in a two-dimensional pattern along the X and Y axis called a Lissajous figure (see Figure 6.6). An identical signal on both channels results in a 180-degree vertical line representing a central mono signal (see Figure 6.7), while a true stereo signal will give you a more or less random figure that's always moving (see Figure 6.8).

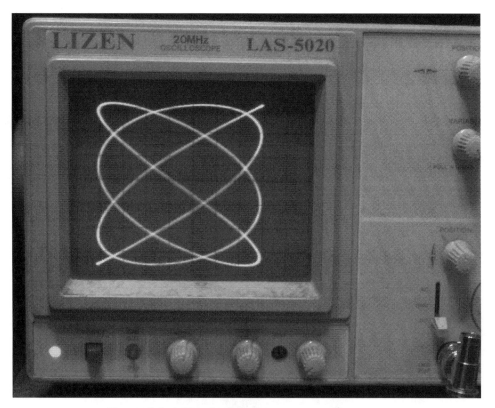

Figure 6.6: A Lissajous figure on an oscilloscope
© 2024 Bobby Owsinski

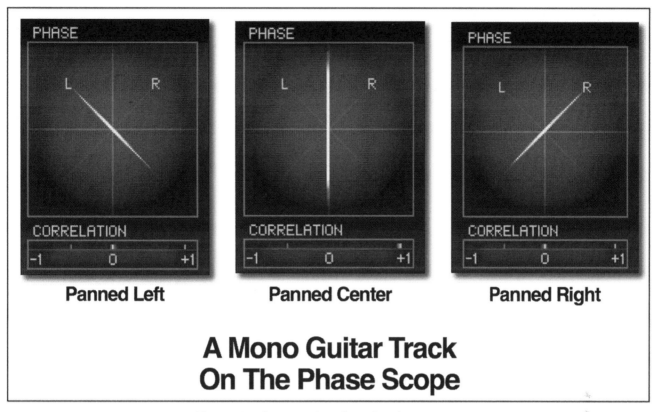

Figure 6.7: A mono signal on the phase scope
© 2024 Bobby Owsinski (Source: IK Multimedia)

Figure 6.8: A stereo signal on the phase scope
© 2024 Bobby Owsinski (Source: IK Multimedia)

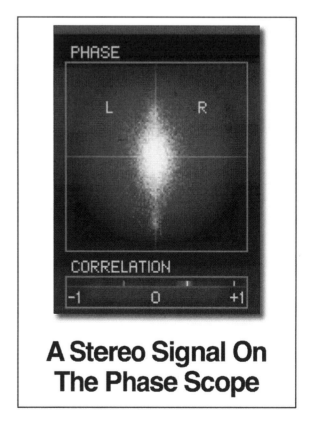

After you watch the phase scope for a while, you'll find that you can instantly tell a lot about the signal as you recognize the different shapes it can take on with different signals. Simpler and nearer sounds, such as mono one-shots or notes and chords, are illustrated by thick, bold-looking, solid lines.

As you widen the stereo image, you'll notice a broader and more string-like pattern. Heavily reverberated or delayed sounds form shapeless images with lots of small dots. The complex arrangements normally found on most records will show all these and everything in between.

The more defined the borders are, the more of the signal is peaking above 0dB. As you can see, the phase scope shows everything from the width, phase, panning, amplitude, and even clipping info in the signal.

Many metering packages that feature a scope function are available, include PSP Stereo Analyzer and the Flux Stereo Tool (which is free), among others.

THE PHASE CORRELATION METER

While the phase scope takes some time to get the hang of, the phase correlation meter is dead simple. Anything drawn toward the +1 side of the meter is in phase, while anything drawn toward the −1 side of the meter is out of phase (see Figure 6.9).

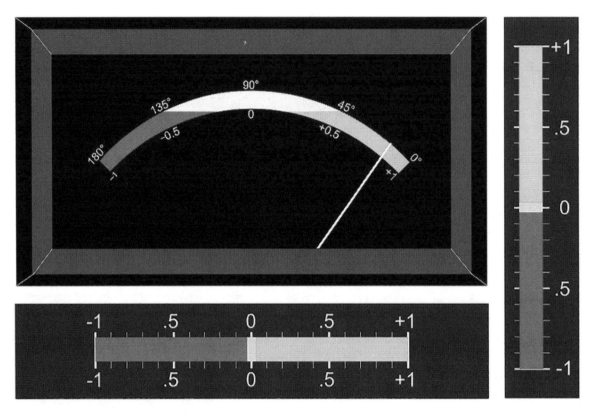

Figure 6.9: Maat phase correlation meters
Courtesy Maat GmbH

In general, any meter readings above 0 on the positive side of the scale have acceptable mono compatibility. A brief readout toward the negative side of the scale isn't necessarily a problem, but if the meter consistently sits in the negative side, it could represent a mono-compatibility issue.

Keep in mind that the wider the stereo mix is, either because of panning or wide stereo reverbs, the more the phase correlation meter will tend to indicate toward the negative side. But as long as the signal stays mostly on the positive, your compatibility should be good to go.

TIP: If the phase correlation meter or phase scope indicates that there might be a mono-compatibility problem, it's important to immediately listen in mono to verify whether it's an issue and whether the track is acceptable.

In the event that an out-of-phase condition is verified, occasionally flipping the phase of one channel can fix it, but sometimes a remix may be the only answer.

Examples include the phase correlation sections of the Flux Stereo Tool, MAAT 2BusControl, and Logic Pro, among others.

THE SPECTRUM ANALYZER

The spectrum analyzer, also known as a *real-time analyzer*, is an excellent tool for determining the frequency balance of your program by displaying it in octave or sub-octave portions (see Figure 6.10). It's especially effective for identifying frequencies that are too hot and for fine-tuning the low end.

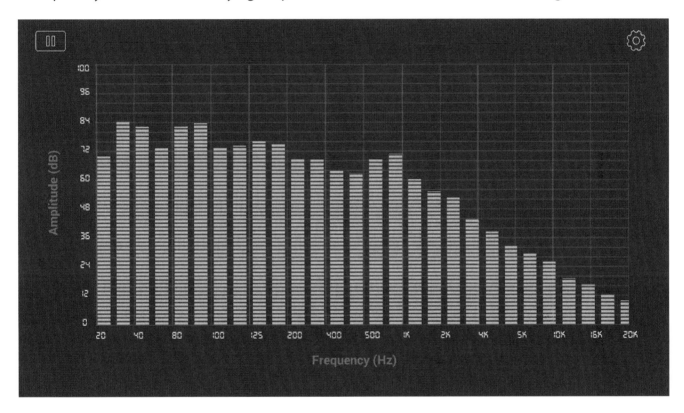

Figure 6.10: A spectrum analyzer
© 2024 Bobby Owsinski (Source: IK Multimedia)

Contrary to what you might think, the goal when using the analyzer is not to aim for a completely flat response. The deep bass (below 40Hz) and the ultra-highs (above 10kHz) almost always have less energy compared to the other frequencies.

It's useful to analyze songs, CDs, mixes, or any audio you think sounds great to understand how they appear on the analyzer.

Keep in mind that your mastering project will probably look different from your reference, as each song is unique. However, if the song is in the same genre, it might be fairly close by the time you've completed your mastering.

The precision of a spectrum analyzer is determined by how many bands the audio spectrum is split into. One-octave analyzers can provide an overall picture of the frequency response, but 1/3- and 1/6-octave versions offer higher resolution into what's happening frequency-wise within the program and are normally used in mastering.

TIP: Most spectrum analyzers allow you to adjust the response time, typically from very fast to a slower average over up to 30 seconds, to provide a clearer view of frequency response over time.

Spectrum analyzers are now part of many EQ plugins, included the Waves F6 RTA and FabFilter's Pro-Q3, and are also included in metering packages by T-RackS and iZotope Ozone, among others. It can also be available as a plugin like True Audio TrueRTA.

The Spectrogram Meter

We're used to seeing a waveform in our DAW displaying a signal's amplitude over time, but that representation is limited in that it fails to show the frequency response of the signal as well. A spectrogram shows those frequencies in a signal over time, with the amplitude represented in varying brightness and color.

In the spectrogram view, the vertical axis shows frequency (in Hertz), the horizontal axis represents time (similar to a waveform display), and amplitude is shown by brightness (see Figure 6.11).

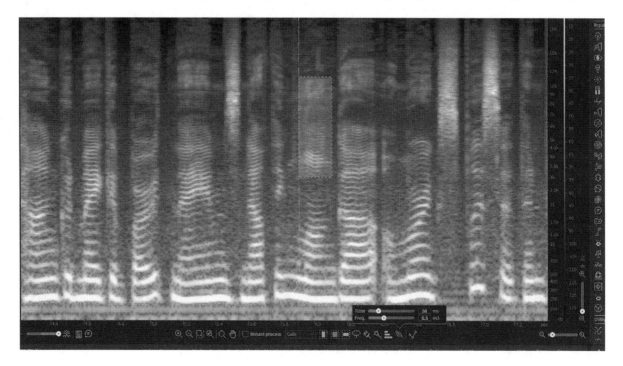

Figure 6.11: iZotope RX spectrogram display
Courtesy iZotope

As audio scrolls from top to bottom or left to right over time, frequencies and their levels are represented by color and brightness, revealing elements that could remain undetected if we were just confined to listening.

The black background indicates silence, while loud events will appear bright and quiet events will appear darker. The lowest frequencies are at the bottom of the display, while highest at the top. A spectrogram display is very important in audio editing since it can clearly display problems that need to be fixed. For instance, you can clearly see:

- Hum
- Buzz
- Hiss and other broadband noise
- Clicks, pops, and other short impulse noises
- Clipping or distortion
- Intermittent noises
- Gaps and drop outs

While mastering engineers usually don't need to fix these issues in recently produced music, the spectrogram can be a lifesaver when working on legacy or archival material.

THE DYNAMIC RANGE METER

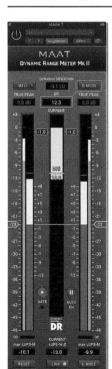

The dynamic range meter is very similar to a peak meter, but it adds the additional function of measuring the dynamic range of an audio signal. *Dynamic range* refers to the degree of variation between the lowest and highest volume levels within a piece of music.

Music with a low value, like DR3, has only a 3dB variation between those levels, indicating heavy compression and little dynamic variation.

A more natural-sounding track might have a DR12 or higher, meaning at least a 12dB difference between the lowest and highest peaks in the song (see Figure 6.12).

Figure 6.12: Maat dynamic range meter showing a value of DR12
Courtesy Maat GmbH

DYNAMIC RANGES OF DIFFERENT GENRES OF MUSIC

Different genres of music sound different at different dynamic range levels. For instance, most acoustic music would sound unpleasant at DR6, but that range might be perfectly acceptable for electronic music. For most pop, rock, R&B, and hip-hop, a DR of 9 might work well, but for jazz, folk, country, or classical music, a DR of at least 12 or higher sounds much better.

Below is a list of average dynamic ranges for different genres of music. As you can see, genres like jazz and classical have a wide dynamic range, while others, like hip-hop and rock, have a much narrower range.

MUSIC GENRE	AVERAGE DYNAMIC RANGE
Hip-Hop	8.38
Rock	8.50
Latin	9.08
Electronic	9.33
Pop	9.60
Reggae	9.64
Funk	9.83
Blues	9.86
Jazz	11.20
Folk, World, Country	11.32
Stage and Screen	14.29
Classical	16.63
Children's	17.03

Dynamic range is one of the most important aspects of mastering, but it's all too often overlooked. As you continue reading this book, keep it in mind as a key tool in shaping the sound of your finished project.

Among the examples of dynamic range meters are the Brainworx bx_meter or the MAAT DR Meter.

THE LUFS METER

The LUFS meter can be confusing to many people creating music, so it deserves a section of its own.

If you're over 30, you might have noticed that the volume levels between TV shows, commercials, and channels are more consistent now, unlike earlier in your life when there were often big jumps in volume between programs. That's because viewers were complaining for years about the fact that there was a dramatic increase in level whenever a commercial aired because it was so compressed compared to the program that you were watching.

Congress set out to address this, and in 2012 adopted the CALM Act, a method to normalize those volume jumps. The European Broadcast Union put similar legislation in place the year before. As a result, the LUFS (Loudness Unit Full Scale) metering system was developed as a standard for measuring audio levels across all delivery mediums.

As anyone who has mixed or mastered knows, the relative volumes of two different songs can be very different even though their levels can look the same on a variety of meters, thanks to the amount of compression and limiting added. This is why the LUFS metering system was developed—to measure the perceived loudness of a program based on how our ears hear it.

LUFS (called LKFS in Europe - they're both the same) measures the perceived loudness of a program by analyzing both transient peaks and the steady-state program level over time using a specialized algorithm. It's different from a normal meter in that it doesn't represent signal level – it measures how loud we perceive an audio program to be because it takes into account the Fletcher-Munson curve, or the tonal balance of human perception.

Typical LUFS Metering

There can be as many as five measurements taken by a LUFS meter. They are:

- **Momentary Loudness** - LUFS Momentary shows the current loudness level just like an RMS meter, except that it takes into account the Fletcher-Munson curve. Applying a slight high-frequency boost provides a reading that's based more on human hearing. Also, it looks at the last 400ms of the signal so that the reading is smoother than an RMS meter.

- **Short-Term Loudness** - Short-term looks at the last 3 seconds of a signal and shows the average loudness level during that time. If the maximum short-term loudness is less than about 6LU below the true peak level (more on this below), that could be an indication that you're compressing more than you need to.

- **Integrated Loudness** - Integrated loudness looks at the average level from beginning to end of the entire program, which could be the duration of a song, commercial, television program or feature film.

- **True Peak** - True Peak differs from Sample Peak in that true peak looks at what occurs between the audio samples. In other words, it's a more accurate peak meter. This meter is 4 times faster than a normal peak meter, taking 192,000 measurements per second. It's a good idea to keep your true peaks under -1dBFS, which leaves 1dB of headroom. Be aware that some platforms such as Spotify, recommend -2dBFS, so you'll need to allow this much headroom if you want to optimize for that platform.

- **Loudness Range (LRA)** - LRA is a statistical method used to measure the variation of loudness across an entire program, where, like with the dynamic range meter, lower LRA values represent lower program dynamics. In this way, the LRA measurement can show whether a program has a continuously constant loudness (low LRA values) or a high variation between low and high level components (high LRA values).

TIP: Loudness Range is meant for dialog and is an unreliable method for measuring the dynamic range of music.

It should be noted that "loudness range" specifically refers to the long-term difference in overall loudness between different sections of a track, while "dynamic range" refers to the short-term difference between the quietest and loudest parts of a single sound or section within a track (like the transient of a snare drum hit). In other words, they're completely different concepts, although they can work well with each other in determining the ideal dynamic range for a project.

Most audio metering looks at what's happening right now in the moment. While LUFS can do that with its Momentary setting, the Short Term and Integrated settings are most often used to show a more or less average level of the entire program.

There are a number of LUFS meter plugins on the market, including ones from TC Electronic, Waves or other developers that show all four metering processes (see Figure 6.13).

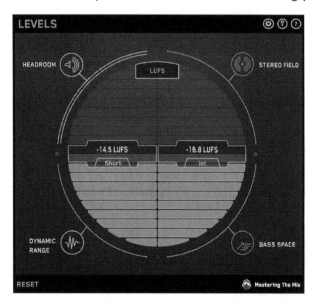

Figure 6.13: LUFS loudness meter
Courtesy Mastering The Mix

Standard LUFS Levels

Looking back at the analog days, mixing level requirements seemed so easy. You aimed for 0 on the VU meter and didn't worry if it bounced over a bit. Of course, under the hood, 0VU could actually be calibrated to different levels, but we usually didn't concern ourselves too much with that as long as the signal was clean around the 0 mark.

These days there are so many different meter reference calibrations available that it can take some time to settle on one that you feel comfortable with. That said, LUFS looms large when it comes to program delivery signal levels, and that makes for lots of confusion.

Although LUFS was originally intended for broadcast audio delivery, it has a new increased meaning in music production as well. Thanks to the fact that streaming services like Spotify and Apple Music are now normalizing the songs uploaded to their catalogs so the level stays the same from tune to tune, there's no real benefit for compressing a song to within an inch of its life anymore. In fact, less volume and more dynamic range are actually your friend.

As mentioned above, LUFS has worked so well in television that other parts of the audio industry adopted it as well, and that's where the confusion sets in. There are a lot of different standard levels, so let's try to sort this out as we look at some of those standards.

- **-24LUFS.** This is your target if you're doing a final edit for television (-23LKFS in the rest of the world). Out of all the standards, this one is the most serious in that a television network can get its broadcast license revoked if it transmits a program or commercial with a level exceeding the standard. Send in a program with a higher level, and it will be kicked back for a revision to the right level.

- **-16LUFS for gaming.** This isn't a hard and fast rule, but it's what the gaming development community has settled on. Nothing serious will happen if you violate this level (you won't lose your game developers license) so it's okay if it's a little off.

- **-16LUFS for podcasts.** Since Apple Podcasts has been the principle delivery system for podcasts since the beginning, this is the level it settled on and the rest of the industry followed. It's usually not a big deal if it's off a little.

- **-16LUFS - Apple Music** (also the Audio Engineering Society standard). You don't have to deliver this level as they're going to re-encode it anyway.

- **-14LUFS - Spotify, YouTube Music, Tidal.** You don't have to deliver this level as these platforms will re-encode it anyway.

- **-9LUFS or less for music.** There is actually no standard LUFS level for music, but it's been determined that songs mixed to this level are sufficiently loud yet still contain some dynamic range.

LUFS In Music

"But Apple Music's standard is at -16LUFS, Spotify is at -14, YouTube is at -14, and my favorite mastering plugin says -9!", you're probably thinking right now.

Here's the thing, *no matter what LUFS level your master is at, Apple, Spotify, Tidal, Deezer and every other streaming service will normalize it to their standard level anyway.* You're just creating more unnecessary work for yourself as a mastering engineer by exporting to different levels for each streaming service (see Figure 6.14).

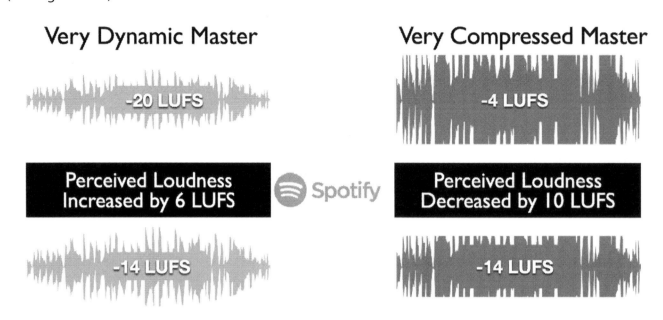

Figure 6.14: How streaming services change the level
© 2024 Bobby Owsinski, All Rights Reserved.

Here what's important though. Since everything is going to be normalized to the same level anyway, there's no reason to crush the master to make it sound louder like in the old days. In fact, *the more dynamic range, the better it will sound* – unless your song is meant for radio or CD. In these cases your song will still be played alongside already loud tracks, so it's best to match their levels in those distribution formats.

Let's put it like this – *A-list mixers and mastering engineers don't worry about LUFS levels, so why should you?* They make it sound as good as they can the same way they've always done (maybe not smashing the level as hard though). The streaming services will adjust the levels to their standards, so it's not something that you have to worry about.

Many music mixers now aim for -9 LUFS as it still allows for some reasonable dynamic range while still sounding fairly loud and competitive. As you approach 0 LUFS, the dynamic range shrinks considerably, although this can be desirable for electronic or pop music, with most songs from those genres averaging around -6 LUFS.

On the other hand, acoustic music benefits greatly from a lower LUFS level of -12 to -14 LUFS. If you analyze hits from 70s and 80s, you'll find that most tend to fall into this range (or even lower), which explains their greater dynamic range. Of course, this was a time way before automatically placing a compressor on your mix buss was the norm like it is today.

You still might want to break out that LUFS meter and find a level that you're comfortable mixing at, but at this point, a dynamic range meter is a better friend.

TIP: Compression and limiting will raise your LUFS level, but use caution—excessive compression can introduce undesirable audio artifacts

---

LUFS doesn't mean much to me. For example, a rock track at -8 LUFS might sound really loud and dense, but an acoustic track at the same -8 LUFS is going to seem way louder than the rock track. It's all about using your ears. If I master something and it sounds good at a certain volume—regardless of the LUFS target—that's where it should stay.

—Pete Lyman

If I master something and it sounds good at a certain volume—regardless of the LUFS target—that's where it should stay.

—Maor Appelbaum

You master for perfect sound and that's it. You don't try to master for a format or it'll bite you in the ass in the end.

—Howie Weinberg

TAKEAWAYS

- A VU meter measures the average signal level, while a peak meter focuses on the signal's highest peaks.

- A VU or RMS meter effectively shows the approximate levels between songs.

- A phase correlation meter or phase scope indicates whether a stereo program has a mono-compatibility problem.

- If the left and right channels are out of phase, the program will sound odd, and instruments panned to the center (like lead vocals and solos) may disappear if the stereo signal is collapsed to mono.

- A spectrum analyzer is a great tool for assessing the frequency balance of your program in octave or sub-octave bands.

- The dynamic range meter displays the variation between the lowest and highest volume levels in a piece of music.

- Music with a low dynamic range like a DR3 indicates there's a lot of compression being used so there isn't much variation in the dynamics of the program.

- The acceptable dynamic range varies by music genre.

- LUFS, Loudness Units Full Scale (LKFS in Europe), measures perceived loudness by analyzing both transient peaks and steady-state levels over time with a specialized algorithm.

- Loudness Range measures the long-term differences in level between sections, while Dynamic Range measures the short-term differences between the loudest and quietest levels in the song.

- The LUFS level of a program is critically important for television program delivery, but only needs to be observed by the final editor.

- Mastering engineers typically disregard LUFS since streaming platforms re-encode all submitted audio.

For Your Best Mix Ever!

Add The Top Selling Book On Mixing For More Than 2 Decades

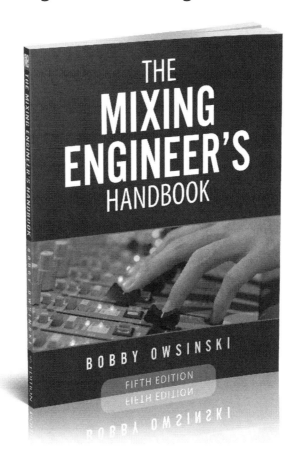

Get 14% off when you order it at bobbyowsinski.com/handbook

(Sorry, due to high shipping costs this offer is only available for customers in the United States, but you can still get it on Amazon)

7

MASTERING TECHNIQUES

Now that you've seen the basic philosophy of mastering, let's tackle the creative aspects. The actual mechanics of mastering can be broken down into several functions, namely maximizing the level of one or more songs, adjusting the frequency balance if necessary, performing any editing, adding fades and spreads, and inserting PQ codes, ISRC codes, and metadata.

What really separates the upper-echelon mastering engineers from the rest is their ability to make any kind of music as big and loud and tonally balanced as possible, while having the taste to know how far to take those operations. On the other hand, the mastering related DAW functions that they use are somewhat mechanical, and don't usually get the same amount of attention as the more creative functions.

We'll look at all of those techniques in this chapter, but first let's look at the basic approach used by most pro mastering engineers.

THE BASIC MASTERING TECHNIQUE

If you were to ask several of the best mastering engineers about their general approach to mastering, you'd get mostly the same answer.

1. Listen to all the tracks. If you're listening to a collection of tracks such as an album, the first thing to do is listen to brief durations of each song (10 to 20 seconds should be enough) to find out which tracks are louder than the others, which ones are mixed better, and which ones have better frequency balances.

By doing this, you can tell which songs sound similar and which ones stand out. Inevitably, unless you're working on a compilation album where all the songs were mixed by different production teams, most of the tracks will have a similar feel to them, and these are the ones to start with.

After you feel pretty good about how these feel, you'll find it easier to make the outliers match the majority, rather than the other way around.

2. Listen to the mix as a whole, instead of hearing the individual parts. Don't listen like a mixer, don't listen like an arranger, and don't listen like a songwriter. Good mastering engineers have the ability to divorce themselves from the inner workings of the song and hear it as a whole, just like the listening public does.

3. Find the most important element. On most modern radio-oriented songs, the vocal is the most important element, unless the song is an instrumental. That means one of your jobs is trying to make sure that the vocal can be distinguished clearly.

4. Have an idea of where you want to go. Before you go twisting parameter controls, try to have an idea of how you'd like the track to sound when you're finished. Ask yourself the following questions:

- Is there a frequency that stands out?
- Are any frequencies missing?
- Is the track punchy enough?
- Is the track loud enough?
- Can you hear the lead element distinctly?

5. Raise the level of the track first. Unless you're extremely confident in your ability to hear a wide frequency spectrum on your monitors (especially the low end), focus on raising the volume rather than EQing. If you feel that you must EQ, refer to the section on EQing later in the chapter.

6. Adjust the song levels so they match. One of the most important jobs in mastering is to take a collection of songs, like an album, and make sure each has the same relative level. Remember that you want to be sure that all the songs sound about the same level at their loudest peaks. Do this by listening back and forth to all the songs and making small adjustments in level as necessary.

> *I'll usually start with the single or the most impactful song, and I'll set it where I can get it as loud as I can before it starts to fall apart and then bring it back 10 percent or so.*
> —Ryan Schwabe

CREATING A LOUD MASTER

The amount of perceived audio volume, or level, without distortion (whether on an audio file, CD, vinyl record, or any future audio-delivery method) is something many top mastering engineers take pride in.

Note the phrase "without distortion," as that is indeed the trick: making the music as loud as needed (and competitive with other products on the market as a result) while still sounding natural.

Keep in mind that this typically applies to modern pop, rock, R&B, electronic, and urban genres, rather than classical or jazz, where listeners tend to prefer a wider dynamic range and maximum loudness isn't a primary concern.

Competitive Level

The volume wars we experience today actually began back in the vinyl era of the 1950s, when it was discovered that if a record played louder than others on the radio, listeners would perceive it as better-sounding and it was more likely to become a hit.

Since then, it's been the charge of mastering engineers to make any song intended for radio as loud as possible in whatever way they can.

This also applies to situations other than the radio, like with a CD changer or streaming music playlist. Most artists, producers, and labels certainly don't want one of their releases to play at a quieter level than their competitor's because of the perception (and not necessarily the truth) that it won't sound as good if it's not as loud.

The limitation of how loud a "record" (we'll use this term generically) can sound is determined by the medium used to deliver it to the consumer. In the days of vinyl records, if a mix was too loud, the stylus would vibrate so much that it would lift right out of the grooves and the record would skip.

When mixing too hot to analog tape, the sound would begin to softly distort and the high frequencies would diminish (although many engineers and artists actually like this effect). When digital audio and CDs came along, any attempt to mix beyond 0dBFS would result in terrible distortion as a result of digital overs (nobody likes this effect).

As you can see, trying to squeeze every ounce of level out of the track is a lot harder than it seems, and that's where the art of mastering comes in.

Level Technique #1: The Compressor-Limiter Tandem

The bulk of the audio-level work today is done by a combination of two of the mastering engineer's primary tools: the compressor and the limiter (see Figure 7.1). The compressor is used to control and increase the level of the source audio, while the limiter handles the instantaneous peaks.

Remember that the sound of both the compressor and the limiter will affect the final audio quality, especially if you push them hard. Here's how you raise the level:

- Set the master level on the limiter to −0.1dB to contain the peaks and avoid digital overs.

- Set a compressor at a ratio of around 2:1 or 3:1 to increase the perceived level. Generally, the trick with compression in mastering is to use either a slow release time or one that's timed to the drums, and less (usually way less) than 3dB of compression.

- Adjust the attack time to allow the desired amount of transients through. In general, the slower the attack time, the punchier the sound.

- Set the release time to avoid hearing any pumping. Time it to the track to keep it punchy-sounding. Set it to slow to keep it smooth-sounding.

- Raise the level of the program to the desired level by increasing the compressor's Output control.

Remember that the less limiting you add, the better it will usually sound. Most of the gain and punch comes from the compressor.

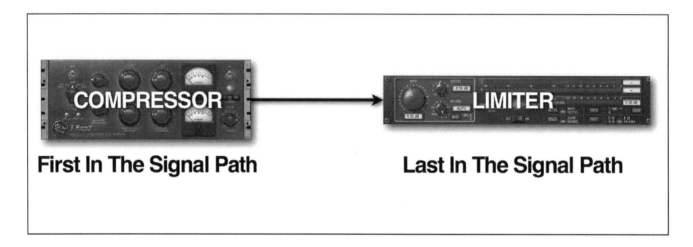

Figure 7.1: The compressor-limiter tandem
© 2024 Bobby Owsinski (Source: IK Multimedia, Universal Audio)

TIP: Use a multi-band compressor and/or limiter to increase the level without hearing as much of the side effects from compression or limiting.

Level Technique #2: Multi-Compressor Packages

Some mastering engineers dislike the sound of limiters so much that they'll go to great lengths to avoid using them. One of the ways this is possible without resulting in the red overload LED continually lighting is by using multiple compressors instead.

In this case, each compressor would be different and would use slightly different compression ratios in order to exert the same kind of control over the signal while increasing the level (see Figure 7.2).

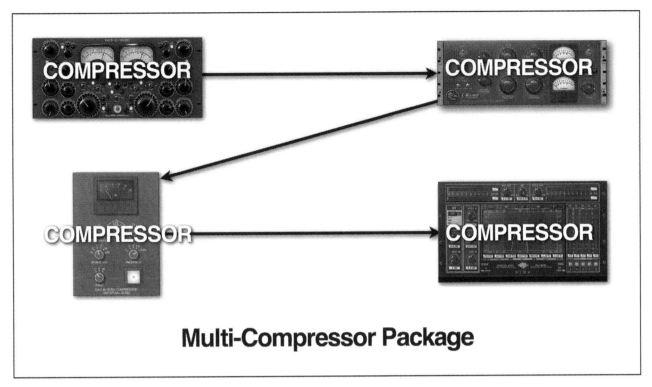

Figure 7.2: Multi-compressor packages
© 2024 Bobby Owsinski (Source: IK Multimedia, Universal Audio)

Advanced Level Techniques

Another way to achieve a high signal level with a minimum of limiting is to insert the limiter only on the sections of the songs with peaks. This involves identifying the brief moments where peaks are strong enough to trigger the limiter, then either automating the limiter to engage at those points or editing and processing those sections separately.

> *I try not to use a limiter if I'm not hearing any distortion. When I run into situations when I need to use a limiter, I'll just use it only on the portions of the song that need it. What I do is go in and slice up a song and just smooth out only the rough edges. I can put up to 300 tiny edits in one song if need be. If it's an open track that has to be loud, I'll just cut all the tiny pieces that need limiting and limit only those. That way those sections go by so fast that your ear can't hear it as it flies by. It gets rid of any overload crackles and keeps the kick hitting hard. It's time-consuming, but I don't mind doing it if it comes out better. It actually goes a lot faster than you think once you have an ear for what to listen for.*
>
> —Gene Grimaldi of Oasis Mastering

The Effects Of Hypercompression

Over the years, it's become increasingly easy to achieve a hotter perceived level, largely due to advancements in digital technology that have led to better, more effective limiters.

Today's digital "look ahead" limiters make it easy to set a maximum level (usually at —0.1dBFS or even —.01dBFS) and never worry about digital overs and distortion again, but this can come at a great cost in audio quality, depending on the situation.

Too much buss compression or over-limiting when either mixing or mastering results in what's become known as *hypercompression*. Hypercompression is to be avoided at all costs because:

- It can't be undone later.

- It can suck the life out of a song, making it sound weaker rather than punchier.

- Streaming platforms using lossy codecs struggle to encode hypercompressed material, often introducing unwanted artifacts.

- It causes listener fatigue, making listeners less inclined to replay your track.

- A hypercompressed track can actually sound worse over the radio because of how it interacts with the broadcast processors at the station.

- A hypercompressed track has little or no dynamics, leaving it loud but lifeless and unexciting. On a DAW, it's a constant waveform that fills up the DAW region. Figure 7.3 shows how the levels have changed on recordings over the years.

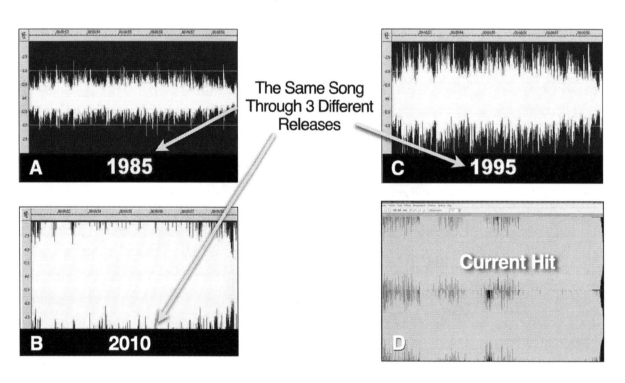

Figure 7.3: From very little compression to hypercompression
© 2024 Bobby Owsinski (Source: Audacity)

This practice has come under fire since we've just about hit the loudness limit, thanks to the digital environment we're now in. Still, both mixing and mastering engineers try to cram more and more level onto the file, only to find that they end up with either a distorted or an over-compressed product.

Mastering Engineer's Handbook - 5th edition | 82

While this might be the sound that the producer or artist is aiming for, it does violate the mastering engineer's unwritten code of keeping things as natural-sounding as possible.

> *When digital first came out, people knew that every time the light when into the red that you were clipping, and that hasn't changed. We're all afraid of the "over" levels, so people started inventing these digital domain compressors where you could just start cranking the level up. I always tell people, "Thank God these things weren't invented when the Beatles were around, because for sure they would've put it on their music and would've destroyed its longevity." I'm totally convinced that over-compression destroys the longevity of a piece. Now when someone's insisting on hot levels where it's not really appropriate, I find I can barely make it through the mastering session. I suppose that's well and good when it's a single for radio, but when you give that treatment to an entire album's worth of material, it's just exhausting. It's a very unnatural situation. Never in the history of mankind has man listened to such compressed music as we listen to now.*
>
> —Bob Ludwig

Competitive Level Isn't What It Used To Be

While everyone still wants a hot mix, making it overly hot is gradually becoming a thing of the past in today's online streaming world.

All online music services, like Apple Music and Spotify, now normalize uploaded content so that it plays at the exact same level, regardless of how loud or quiet the original master was.

As a result, there's no longer a good reason to make the master levels excessively hot. In fact, that's proven to actually be counterproductive. *A song with more dynamic range and a lower level will end up sounding better than the louder one after the service's normalization process.*

Mastering engineers everywhere (as well as mixing engineers) will jump for joy as sanity returns to mixing levels and the "volume wars" finally end. We're not quite there yet, but progress is definitely being made.

Setting The Compressor

The key to getting the most out of a compressor are the *Attack* and *Release* (sometimes called *Recovery*) parameter controls, which have a tremendous overall effect on a mix and therefore are important to understand.

Generally speaking, transient response and percussive sounds are affected by the *Attack* control setting. The *Release* control determines how long it takes for the gain to drop back below the threshold and return to zero gain reduction.

In a typical pop-style mix, a fast *Attack* setting will react to the drums and reduce the overall gain. If the *Release* is set very fast, then the gain will return to normal quickly but can have an audible side effect of reducing some of the overall program level and attack of the drums in the mix.

With a slower *Release* setting, the gain changes may cause the drums to produce an effect called pumping, where the mix level rises and falls noticeably. Each time a dominant instrument starts or stops, it "pumps" the mix level up and down.

Compressors that work best on full program material generally have very smooth release curves and slow release times to minimize this pumping effect.

TIP: To minimize the pumping effect, avoid very fast release times.

Compression Tips And Tricks

Adjusting the *Attack* and *Release* controls on the mastering compressor and/or limiter can have a surprising effect on the program sound.

- Slower *Release* settings generally make gain changes less noticeable but lower the perceived volume.

- A slow *Attack* setting will tend to ignore drums and other fast signals but will still react to the vocals and bass.

- A slow *Attack* setting may also allow a transient to overload the next plugin in the signal chain.

- Compressor gain changes triggered by drum hits can reduce the level of vocals and bass, causing noticeable volume changes throughout the program.

- Usually only the fastest *Attack* and *Release* settings can make the sound "pump."

- The more the level meter bounces, the more likely it is that the compression is audible.

- Quiet passages that are too loud and noisy are usually a giveaway that you are seriously over-compressing.

Don't just set those *Attack* and *Release* controls to the default setting and forget about them. They can make a significant difference on your final mastered sound.

Setting The Limiter

Most digital limiters used in mastering are set as brick-wall limiters. This means that no matter what happens, the signal will not exceed a certain predetermined level, preventing any digital overs.

Thanks to the latest generation of digital limiters, it's easier than ever to achieve louder levels, thanks to more efficient peak control. This is primarily due to the look-ahead function that almost all modern digital limiters now feature.

Look-ahead delays the signal slightly (by about 2 milliseconds), allowing the limiter to anticipate and process peaks before they occur. Since the threshold cannot be overshot, the limiter is considered a brick-wall limiter.

Analog limiters don't work nearly as well in this situation because they can't predict the input signal like a digital limiter with look-ahead can.

Most limiters intended for mastering will have a *Ceiling* or *Output* control to set the maximum output level (see Figure 7.4). For the absolute loudest CD levels, this is typically set around -0.1dB or even -0.01dB. For online delivery, the level is usually set lower, at -1dB or less, because most encoders produce a slightly higher output.

> *My guideline is to keep the loudest sections of a track no louder than -10 LUFS, with true peaks at -1 dB to leave headroom for encoding to lossy formats like MP3. You want to ensure the loudest parts sound good first, then balance everything else musically. Ideally, the mix is already balanced, but if not, you can use a little gentle automation to adjust the quieter sections.*
>
> —Ian Shepherd

For a vinyl record, the level may be set even lower. A *Threshold* parameter then controls the amount of limiting that will take place, which is then measured by the gain reduction meter.

Ideally, no more than 2 to 3dB of limiting should take place in order to keep the limiting less audible, although this amount might be even less for a vinyl master.

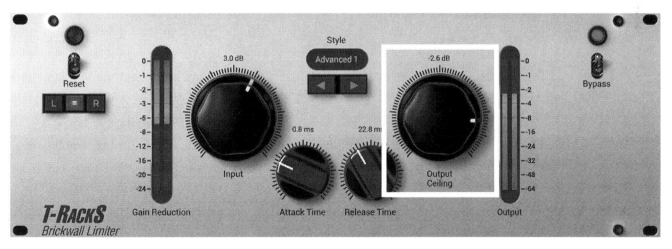

Figure 7.4: A digital brickwall limiter's Ceiling control
© 2024 Bobby Owsinski (Source: IK Multimedia)

Most digital limiters only feature a *Release* control, as the *Attack* control is unnecessary due to the look-ahead function. Nearly all digital limiters include an *Auto-Release* function, which automatically sets the release time based on the audio program being processed.

This is generally a safe selection, although manually setting the *Release* can be effective in keeping the audio punchy.

TIP: *A release time that's too long can dull the sound and cause the mix to lose its punch.*

Using Multi-Band Compressors And Limiters

As mentioned earlier, multi-band compressors and limiters are highly effective at increasing level with fewer audible side effects than single-band units. This is because they operate on small portions of the frequency spectrum at a time, leaving the other frequencies relatively unaffected (see Figure 7.5).

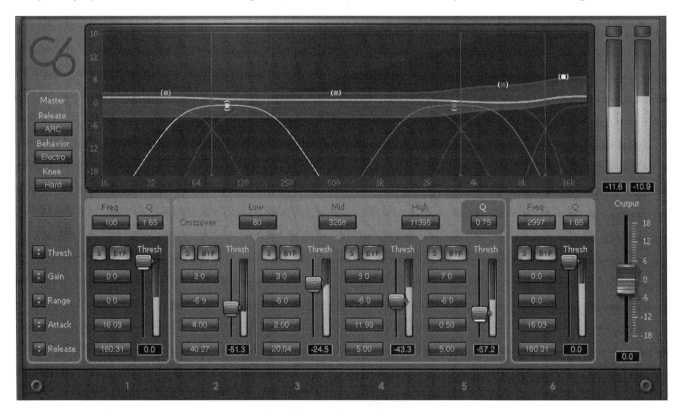

Figure 7.5: The Waves C6 multiband compressor
Courtesy Waves

However, to minimize side effects, a few precautions should be observed:

- Use the same compression ratio across all bands, as differing ratios can create an unnatural sound.

- Apply roughly the same amount of gain reduction to each band to avoid altering the overall mix balance too much.

- Excessive compression or limiting in one band often indicates that the band has been over-EQed. In this case, either reduce the EQ (if it was applied before the compressor or limiter), cut the problematic frequency, or suggest that the mix be redone.

Limiting the program comes at a cost because the limiter does change its sound, softening transients on the drums and other percussive instruments, and reducing the punch of a track. For this reason, most mastering engineers aim to avoid excessive limiting or, ideally, use none at all when possible..

Reducing Sibilance With A De-Esser

Sibilance is a short burst of high-frequency energy that over-emphasizes "S" sounds, and reducing it requires a special type of compressor called a de-esser (see Figure 7.6).

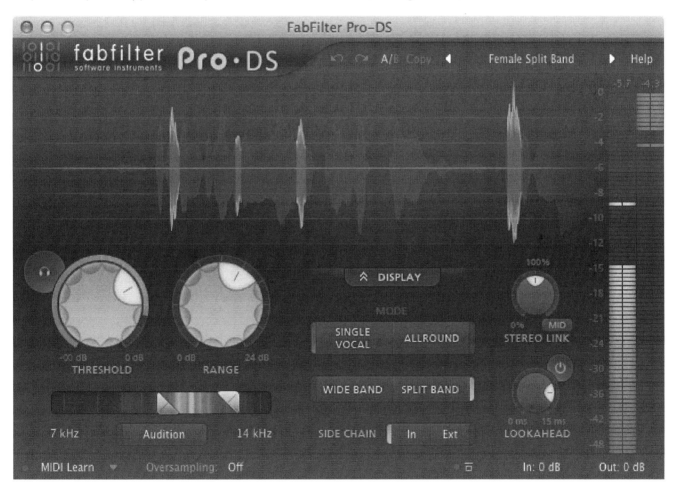

Figure 7.6: FabFilter Pro-DS De-esser
Courtesy FabFilter

Sibilance often occurs in a mix when heavy compression is applied to either the vocal or the mix buss, or EQ is added to make the vocal rise above the mix. To use a de-esser, follow these steps:

1. Insert the de-esser.
2. Lower the *Threshold* control until the sibilance is decreased but you can still hear the S's. If you can't hear them, then you've lowered the *Threshold* too far.
3. Sweep through the available frequencies using the *Frequency* control to find the most problematic spot, then adjust the *Threshold* until the "S" sounds are more natural.
4. Use the *Listen* feature to determine the exact sibilance frequency.

TIP: When using the Listen feature, remember that the audio you hear is not part of the signal path, but just the sidechain. Don't forget to disengage Listen after you've found the correct frequencies.

FREQUENCY BALANCE TECHNIQUES

EQing is often where engineers run into trouble when mastering their own mixes. There's a tendency to overcompensate with EQ, especially by adding too much low end, which can completely ruin the frequency balance. Fortunately, there are some rules to help avoid this.

Rule 1. Listen to other CDs or high-resolution files that you like first, *before you touch an EQ parameter*. The more reference tracks you listen to, the better. You need a reference point for comparison; otherwise, you're likely to overcompensate.

Rule 2. A little EQ goes a long way. If you feel that you need to add more than 2 or 3dB, you might be better off mixing the song again!

During mastering, EQing is almost always in small increments of anywhere from tenths of a dB to 2 or 3dB at the very most.

Seriously though, if you find yourself adding a lot of EQ, go back and remix. That's what the pros do. It's not uncommon for a mastering engineer to communicate with a mixer, point out what's off, and suggest a remix.

Rule 3. Keep comparing the EQ'd version to the original version. The goal of mastering is to make the song or program sound better with EQ, not worse.

Don't fall into the trap where you think it sounds better just because it's louder. To accurately judge what you're hearing, keep the levels of the EQ'd and pre-EQ'd tracks as close as possible.

That's why an app like Adaptr Audio Metric AB or NUGEN Audio AB Assist can be really useful for mastering. They have an A/B function that lets you compensate for level increases so you can accurately determine if you're improving the sound.

Rule 4. Compare the song you're working on with the other songs in the project. The idea is to get them all to sound somewhat the same.

It's common for mixes to sound different from song to song, even if they're done by the same mixer with the same gear in the same studio. Your job is to make the listener feel like all the songs were done on the same day in the same way. They should sound as close to each other as possible, or at least close enough that none stand out significantly.

TIP: Even if the songs don't match your best-sounding reference track, your mastering will still be considered "professional" if you can make all the songs in the project sound consistent in tone and volume!

My approach to mastering, especially with EQ, is based on a theory: most of the time, your first impression is the right one—that's the lightning in a bottle. If you spend too much time obsessing over something, you'll lose that initial spark.

—Pete Lyman

If it needs nothing, don't do anything. I think that you're not doing a service by adding something it doesn't need. I don't make the stew, I season it. If the stew needs no seasoning, then that's what you have to do, because if you add salt when it doesn't need any, you've ruined it.

—Doug Sax

THE MASTERING SIGNAL PATH

The placement of processors in the signal path can significantly impact the mastering process. Here are a few possible setups.

The Basic Mastering Signal Chain

In its simplest form, the mastering signal chain consists of three elements: an equalizer followed by a compressor, which is then followed by a limiter (see Figure 7.7).

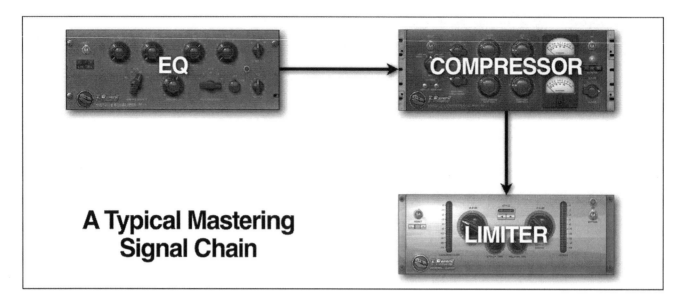

Figure 7.7: A basic mastering signal chain
© 2024 Bobby Owsinski (Source: IK Multimedia)

This arrangement is a holdover from the analog days when this processor order produced the best sound quality when transferring a program into a workstation from an analog source like tape. Often, it wasn't possible to get enough gain from the EQ's output stage without introducing distortion (especially if one band was pushed hard), whereas the compressor was typically designed to add plenty of clean gain when necessary.

That's not as much of an issue in the digital domain, especially when applying small 1 and 2dB increments of boost. Still, placing the EQ first in the chain remains a popular choice. *The downside is that whatever frequency is boosted by the EQ may be the first to trigger the compressor,* potentially causing unexpected and unpleasant results.

This is why it's often fine to reverse the order, placing the compressor first and the EQ second, especially if large amounts of EQ are needed.

As a result, a general rule of thumb for compressor/EQ order goes like this:

- If you need to apply significant EQ, place the EQ after the compressor.
- If you're using a lot of compression, place the compressor after the EQ.

The limiter is always the last in the chain, no matter how many other devices you add and in which order, because that's what stops any digital overs from occurring.

> *I usually use two limiters. The first Limiter is getting the final loudness, and then I'm doing some processing in between, and then a last limiter catches any peaks,*
>
> —Ryan Schwabe

An Advanced Signal Chain

Sometimes the simple setup described above isn't enough for a particular mastering job. Anytime you feel the need to apply extreme processing (such as 10dB of EQ or 6dB of gain reduction), it's often better to use small amounts from multiple processors to keep the signal clean, smooth, and punchy.

Here's the approach that many pro mastering engineers take, opting to use small amounts of processing from a number of processors (see Figure 7.8). Once again, the limiter is at the end of the chain.

> *For me, the standard chain is EQ, compression, and limiting. If you have one good unit handling each of those stages, you can get great results on about 80% of the material. Sometimes you might want to use multiple stages of compression, swap out the limiter for a different sound, or use clipping along with limiting—but the core chain is usually the same: EQ, compression, and limiting.*
>
> —Ian Shepherd

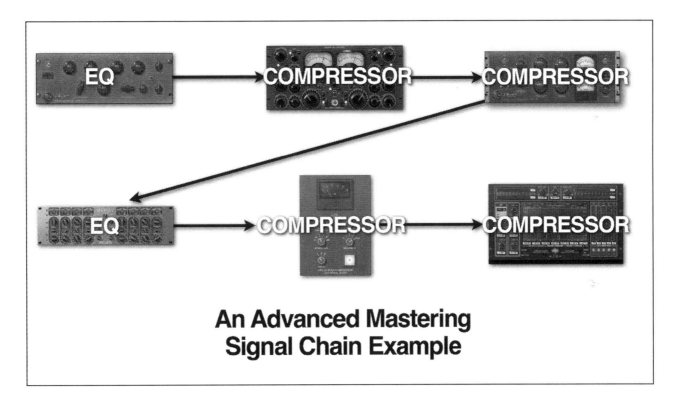

Figure 7.8: An advanced signal chain using multiple processors
© 2024 Bobby Owsinski (Source: IK Multimedia, Universal Audio)

Parallel Processing

While you can keep adding processors to the signal path, each doing a little processing instead of one doing a lot, remember that each processor is influenced by the previous one in the chain. Sometimes, a better solution is parallel processing.

In parallel processing, the main signal is split into two separate processing chains and then recombined before entering the limiter. T-RackS is a mastering application well-suited for this, offering multiple parallel slots in its first four processor positions (see Figure 7.9).

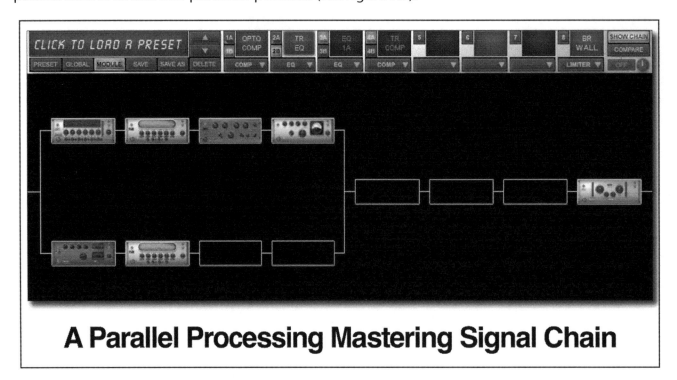

Figure 7.9: A parallel-processing signal chain in T-RackS 6
© 2024 Bobby Owsinski (Source: IK Multimedia)

The signal path is critical to mastering success. Whether it's a simple three-processor chain or something much more complex, always be sure that the limiter is the last processor in the path.

Just as a reference point, most major mastering facilities have both analog and digital signal paths, since so many of the tools and source materials exist in both domains. That being said, the signal path is kept as short as possible, with unnecessary gear removed to avoid inadvertently affecting audio quality.

ADDING EFFECTS

While mastering engineers have occasionally been asked to add effects over the years, this has become much more common recently. This is partly due to the widespread use of digital audio workstations, where poorly executed fades or song tails being cut off during file preparation are more frequent issues.

Additionally, many artists and producers are sometimes shocked to find that the amount of reverb is noticeably less than they remembered during the mix.

Most mastering engineers prefer to add effects in the digital domain, both for ease of use and sonic quality. Reverb plugins, such as Audio Ease Altiverb or LiquidSonics Seventh Heaven, are commonly chosen for this purpose.

A lot of people assemble mixes on Pro Tools, and they don't listen to it carefully enough when they're compiling their mix, so they actually cut off the tails of their own mixes. You can't believe how often that happens. A lot of times we'll use a little reverb to just fade out their chopped-off endings and extend it naturally. I do a fair amount of classical-music mastering, and very often a little bit of reverb is needed on those projects, too. Sometimes if there's an edit that for some reason just won't work, you can smear it with a bit of echo at the right point and get past it. Sometimes mixes come in that are just dry as a bone, and a small amount of judicious reverb can really help that out.

—Bob Ludwig

EDITING TECHNIQUES FOR MASTERING

Today's mastering engineer is called on to do several types of editing that, while similar to what you might do during production, are quite specialized. Here are a few examples.

Inserting Fades

Sometimes a default fade that's added to the beginning or end of a track just doesn't sound natural. Either the fade is too tight and cuts off the head or tail of the track, or the fade itself just doesn't sound smooth-sounding.

If that's what you hear, then now is the time to fix any fades that don't work in the track by adjusting the fade timings.

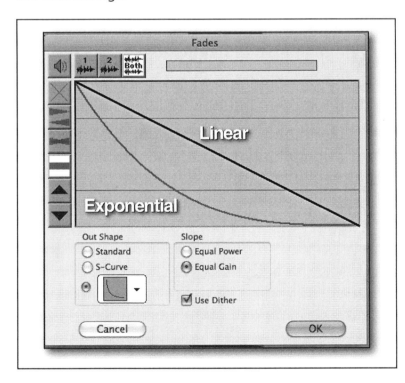

If a fade (particularly a fade-out) feels unnatural, try using an exponential power fade instead of the default (see Figure 7.10).

Figure 7.10: A linear versus an exponential fade
© 2024 Bobby Owsinski (Source: Avid)

TIP: Default fades often don't sound smooth enough, especially in fade-outs. Be ready to adjust the fade by experimenting with other fade types.

---

93 | Mastering Techniques

Eliminating Intro Noise And Count-Offs

Leaving noise or count-offs, such as drumstick clicks on a song intro, is a sure sign of a demo recording and is something that usually no one wants to listen to. The solution is to use a fade-up while ensuring you don't cut off the attack of the song's downbeat.

Clean intros are a sign of a professional mastering job. It only takes a minute, but can make a big difference in how the song is perceived.

Making A "Clean" Master

When a mix contains lyrics that some listeners might find objectionable, the mastering engineer may be asked by the record label to create a "clean" version suitable for radio airplay. This can be done in several ways:

- If a TV track or an instrumental mix is available, edit in a piece everywhere that the objectionable lyric takes place. The vocal will drop out for the duration of the lyric, but the song will continue and it will fly by so fast that most casual listeners may not notice if it's short enough.

- Find a similar instrumental section in the song, and copy and paste it in place of the objectionable lyric, similar to option #1.

- If no alternate track is available, replace the offensive lyric with a 1kHz sine wave. This method stands out and can be effective if the artist wants to highlight that they've been censored. However, it may not work well if there are many instances or long durations, so try to get an instrumental mix from the producer or label if possible.

PARTS PRODUCTION

Although the more high-profile and documented part of mastering takes place in the studio with the mastering engineer, the real bread and butter of the business happens afterwards, during what's known as *production*.

Production is when the various masters are created, verified, and sent to the replicator. While not a very glamorous portion of the business, it's one of the most important nonetheless, since any mistakes here can negate a perfect job done beforehand.

In the past, production was a lot more extensive than it is today. For instance, during the vinyl era, many masters had to be made because a pair (one for each side of the disc) had to be sent to a pressing plant in each area of the country, and overseas if it was a major international release.

When you consider that every master had to be cut separately with exactly the same processing, you can see that the bulk of the mastering work was not in the original rundown, but in the actual making of the masters (which was very lucrative for the mastering house). In fact, many large mastering facilities employed a production engineer just for this purpose, sometimes running a night shift at the studio (a few of the larger mastering houses still do this).

Over the years, parts production has dwindled to the point that we're at today, where digital copies are so easy to produce that the mastering engineer handles them at the end of the session.

The QC process is the most important part of mastering. If the song is good, your job is to make sure it's production-ready. I know that seems boring to some people, but that's what the song needs.
—PETE LYMAN

MULTIPLE MASTERS

A project may require several different masters, depending on marketing plans and the label's policy. These typically include:

- **The CD master.** This is the master used to create the glass master at the pressing plant, which is then used to replicate CDs (see Chapter 12 for more on CDs).

- **The vinyl master.** For vinyl releases, a separate master for each side of the disc is required. Vinyl masters are often made at a lower level and with less compression and limiting.

- **The Apple Digital Masters master.** Since online distribution is now a significant part of an artist's revenue stream, a separate master for Apple Digital Masters may be created (see Chapter 9). This master is typically tweaked with a lower level to optimize fidelity within the available bandwidth.

- **Alternate takes.** Many artists signed to major and large indie labels also supply alternate takes, such as TV or instrumental mixes. A TV mix includes the full mix minus the lead vocal, and allows the artist to sing live against the track when appearing on television. An instrumental mix contains the entire track minus any lead and background vocals.

- **Backup masters.** Most major labels request a backup master for storage in the company vault. Many mastering facilities also create a "house" backup to save time in case a new master is needed later.

Then the alternate versions that generally come with major label projects is extensive. It's like the main version, the TV version, the clean version, the instrumental version, and then the acapella version. There might also be the acapella clean version and the TV clean version as well. So that's sometimes six or seven alternate versions per track. I did the Friday record that had six or seven versions per track. It was a 12 track album so we're talking 90 deliverables of alternates and mains.
—RYAN SCHWABE

MASTERING MUSIC FOR FILM AND TELEVISION

Except on rare occasions, the only audio that gets mastered for film is for the songs intended for the movie, as the film studio or production company usually creates their own underscore, dialogue, and effects.

Even though the film studios create the underscore music, occasionally a recording artist is asked to record or re-record songs specifically for a film. Since the artist often prefers to follow their usual workflow, the songs may be mastered. In these cases, the songs are mastered as usual, but it's important to involve the film's music supervisor to ensure that the deliverables are suitable for the music editor.

The Soundtrack Album

One of the more difficult mastering jobs for any mastering engineer is the soundtrack album for a film. This is a compilation of all the songs from the movie that must be mastered as an album.

The challenge lies in the fact that the songs often come from different artists and producers, recorded in various studios with different musicians. As a result, each song may sound drastically different, and it's the mastering engineer's job to unify them in tone and level, aligning with the original goal of mastering.

There are several approaches to this, but a common method is to start by working on the most difficult song first, then match the tone of each subsequent song to it. After that, the songs are adjusted to match in perceived volume.

Mastering Music For Television

Most of the time, mastered music intended for television is delivered to the post-production facility or video editor, where it's mixed against the video. The video editor determines the correct level relative to the effects and dialogue, similar to how it's done in film.

Even though the federally mandated loudness specification for television is −24LUFS (see the section on LUFS measurement in Chapter 6), you typically don't need to worry about hitting the exact television LUFS level. Just like with music supplied to streaming services, the music you master will be sent to the video editor, and it's their job to ensure the proper LUFS level is applied to the files delivered to the network or streaming service.

> **TIP: Keep in mind that, for both film and television, it's the last person in the production chain who is responsible for setting the proper LUFS level. When it comes to television, that's not the mastering engineer.**

TAKEAWAYS

- Making a group of songs sound like they belong together in tone and level is a major job of the mastering engineer.

- The simplest mastering signal chain consists of a compressor, followed by a limiter.

- If the attack time of a compressor or limiter is set too fast, the transients of the song—especially the drums—will sound dull and weak. A slower attack time results in a punchier sound.

- If the release time is set too fast the track may develop a pumping effect. A slower release time will produce a smoother sound.

- The increase in track level during mastering primarily comes from the compressor.

- Sometimes multiple compressors are used in place of a limiter.

- The limiter *Ceiling* or *Output* control is typically set to around -0.1dB for maximum level without digital overs.

- Using a multi-band compressor and/or limiter can help increase the level while minimizing the side effects of compression or limiting.

- Excessive buss compression or over-limiting during mixing or mastering leads to hypercompression, which can drain the life out of a song.

- For vinyl records, the overall master level is often set considerably lower than the main digital master, sometimes with less compression as well.

- In general, professional mastering engineers make small EQ boosts and cuts across different adjoining frequencies, rather than applying major adjustments to a single frequency.

- An exponential fade often sounds smoother and more natural than a straight linear fade.

- It's not uncommon for a mastering engineer to fix the fade or ending of a song when it's been cut off during export.

- Mastering engineers are rarely concerned with LUFS levels, as masters are typically re-encoded by streaming platforms upon submission.

8

DEDICATED MASTERING PLUGINS

It wasn't that long ago that the first dedicated mastering plugin was introduced. Before then, mixers had to rely on the methods shown in Chapter 7 or send their mixes to a mastering engineer.

Today there's a wide range of mastering plugins that range from very simple to AI-assisted mastering plugins, all aimed at giving you the ability to create the best master possible while keeping your budget down. Let's look at them.

DEDICATED MASTERING PLUGINS

Dedicated mastering plugins range from loudness maximizers, to all-in-one mastering processors, to specific plugins based on top mastering engineers' workflows, all aimed at making the process easier.

Maximizers

Many manufacturers and engineers use the terms 'limiter' and 'maximizer' interchangeably, but there is a distinction between them..

A limiter prevents an audio signal from exceeding a user-defined threshold by attenuating the loudest peaks. This prevents a track from clipping during mixing and mastering, or it can increase the average level of a track without its peaks clipping the mix buss.

A maximizer uses limiting as part of its process, but is specifically designed to increase a track's loudness.

In other words, a limiter sets a ceiling that the loudest peaks can't exceed, while a maximizer pushes the music up towards the ceiling.

Maximizers may also provide parameters that vary the tonal color, enhance transients, and provide true-peak protection, oversampling, and dither (see Chapter 12).

Both limiters and maximizers have very fast attack and release times, which can cause distortion if pushed too hard. Most maximizers minimize this distortion by intelligently adjusting the attack and release times according to the program signal.

Like limiting, maximizing can sometimes negatively affect a mix if overused. Be sure to listen carefully to avoid increased distortion or loss of punch.

Some examples of maximizers include the JST Maximizer, UAD Precision Maximizer, Melda Production MUltraMaximizer, Sonnox Oxford Inflator (see Figure 8.1), and Rob Papen MasterMagic.

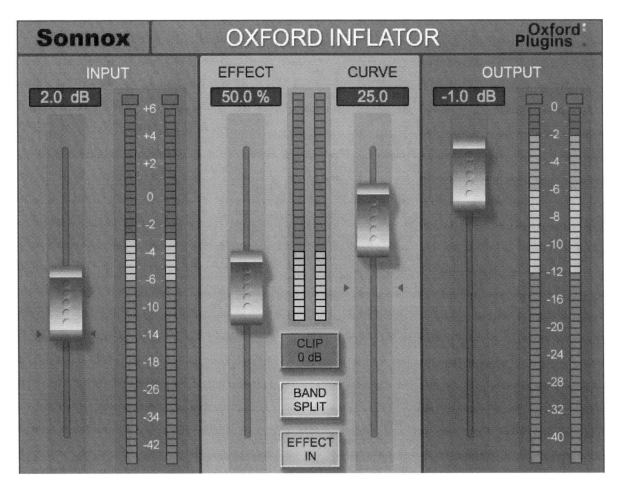

Figure 8.1: Sonnox Oxford Inflator maximizer plugin
Courtesy Sonnox

Comprehensive Processors

While maximizers focus primarily on dynamics processing, comprehensive processors are essentially like having a mastering engineer's console in plugin form.

These typically include an equalizer, compressor, and limiter, along with features like stereo widening, saturation, filters, de-essing, metering, and even maximizing.

While these plugins offer much more power, they take time to learn due to the wide range of available parameters. With all that power at your fingertips, it's easy to overdo it, potentially making your master sound worse than if you stuck with the simple signal path from Chapter 7.

Examples include Brainworx bx-masterdesk (see Figure 8.2) and masterdesk Pro, United Plugins Mastermind, Softube Flow, and Musik Hack MasterPlan (see Figure 8.3).

Figure 8.2: Brainworx bx-masterdesk
Courtesy Brainworx

Figure 8.3: Musik Hack MasterPlan
Courtesy Musik Hack

101 | Dedicated Mastering Plugins

Dedicated Workflows

Dedicated workflows emulate the signal chains of high-profile mastering engineers or, in the case of Abbey Road Studios, the signal chains used on major records in the past

Both the Gavin Lurssen Mastering Console (see Figure 8.4) and the Howie Weinberg Mastering Console plugins offer complete emulations of their respective signal chains

While the Lurssen Console is a little easier to use unless you jump under the hood to access a host of additional parameters, the Howie Weinberg Console provides control over just about every parameter that anyone might ever need for a mastering job.

The Lurssen Console can also operate in a stand-alone desktop mode and be controlled remotely from an iPad.

The Abbey Road TG Mastering Chain plugin (see Figure 8.5) emulates the EMI TG12410 Transfer Console, the original solid-state unit built in-house by EMI engineers in the early '70s. This console has been used on hits from Pink Floyd to Ed Sheeran. It features Input, Tone, Compression/Limiting, Filter/EQ, and Output/Stereo Spread sections that can be arranged in any order.

Figure 8.4: The Lurssen Mastering Console plugin
Courtesy IK Multimedia

Figure 8.5: The Abbey Road TG Mastering Chain
Courtesy Waves

TIP: Workflow mastering plugin presets are particularly effective because they're fine-tuned by the original users of the signal chains.

The purpose of dedicated workflow mastering plugins is to simplify mastering by offering all the necessary tools in a single plugin. Plus, they include the 'special sauce' of the mastering engineer or studio.

> *One problem is that it doesn't give people a chance to learn and actually have the knowledge base that I've been graciously allowed to absorb over time. The tools often make the decisions for you and you don't get to learn much about what is happening.*
>
> —Ryan Schwabe

AI-ASSISTED MASTERING PLUGINS

Mastering plugins can either be a game-changer or a setback, depending on how they're used. The oldest entry in this category by far is iZotope's Ozone, which was first released in 2001. In 2017, the plugin introduced machine learning with its Tonal Balance module (also available as a standalone plugin), and in 2019, it unveiled its latest AI feature, Master Assistant (see Figure 8.6).

Figure 8.6: iZotope Ozone
Courtesy iZotope

Ozone is incredibly powerful, boasting 11 modules (Clarity, Maximizer, Equalizer, Impact, Stabilizer, Imager, Low End Focus, Master Rebalance, Spectral Shaper, Dynamic EQ, Exciter, Dynamics, and Match EQ). But as Spider-Man says, 'With great power comes great responsibility!"

This potent plugin has so many powerful features that it's easy for users to accidentally harm a song rather than improve it if they aren't careful.

Where the AI comes in is with its Master Assistant, which is where Ozone's true potential really lies. As with other AI-assisted audio plugins and their *Learn* buttons, pressing the Master Assistant button will engage the AI and you then play the track.

TIP: For best results with the Master Assistant, play the loudest section of the song for about 30 seconds.

What's brilliant is that there's no need to select a genre profile as Master Assistant automatically recognizes the type of music played. If you think it got it wrong, you can always select a new target genre from the list on the left that pops up.

Master Assistant will then automatically select the processing modules needed and set their parameters as well. You can always manually make adjustments, but this is where users frequently run into trouble by trying to outthink the AI to make the song "sound better."

> **Tip: It's best to upload a reference track into Ozone so it can analyze its settings and apply them to your master. From there, trust the AI!**

Unless you have an excellent playback system with an acoustically-treated room to match, it's unlikely that you'll make the song sound any better, and you run the danger of making it sound worse.

> *In terms of loudness, iZotope's Ozone Maximizer seems to offer the most range I can get out of a single limiter plugin. It allows me to shape how I want the limiter to work. If I want to enhance transients, I can. If I want to smooth the mids and deepen the subs, I can. If I want things to sound flatter, softer, and less exciting, there's also a setting for that. Ozone lets me do a lot of micro dynamic shaping in terms of how it handles loudness, so it's a primary tool for me.*
>
> —Ryan Schwabe

> *I think AI-based plugins are really cool, but it annoys me that they call themselves "mastering" plugins because they're not really mastering. It's more like automated EQ and dynamics balancing. It's like the wizard in Photoshop that automatically makes an image look "fine." A lot of the time it does a reasonable job, but sometimes it's horribly wrong, and other times it's just "take it or leave it."*
>
> —Ian Shepherd

Other Features

One of the cooler features of Ozone is Tonal Balance, which shows you the frequency curves of typically mastered songs in your genre so you can try to match the frequency response (Figure 8.7). Ozone usually does a great job, so minimal tweaking is needed, but if your mix's frequency curve is far outside the suggested range, you may need to go back and remix.

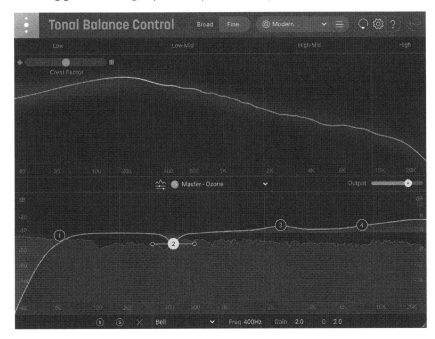

Figure 8.7: iZotope Tonal Balance
Courtesy iZotope

A helpful free add-on tool for Ozone is Audiolens (see Figure 8.8), which lets you load multiple reference tracks and instantly access them from within Ozone. This saves time, as you won't need to search for and load the right track each time you start mastering. The perfect reference is only a click away.

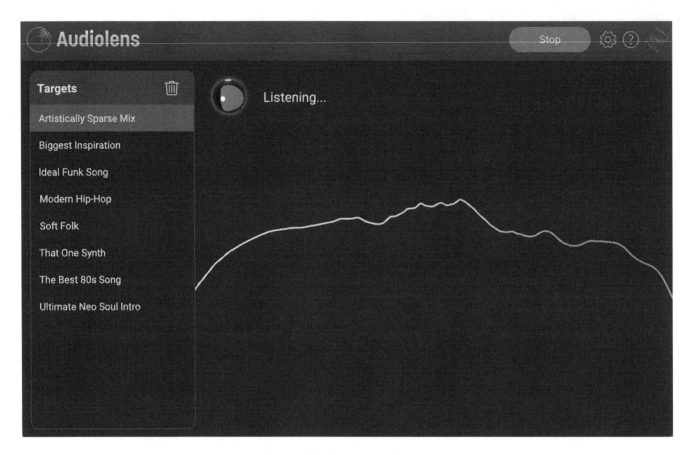

Figure 8.8: iZotope Audiolens
Courtesy iZotope

Other AI-based mastering plugins from Exonic AI Master and LANDR might not be as extensive as Ozone, but the can also be easier to use. While AI Master will enhance and optimize the dynamic range, spectral balance, stereo field and overall level of your song, it offers no manual parameter controls or genre selection.

The LANDR plugin (see Figure 8.9), on the other hand, is the plugin version of their popular online mastering service. LANDR draws on over 10 years of experience with millions of tracks used as training material.

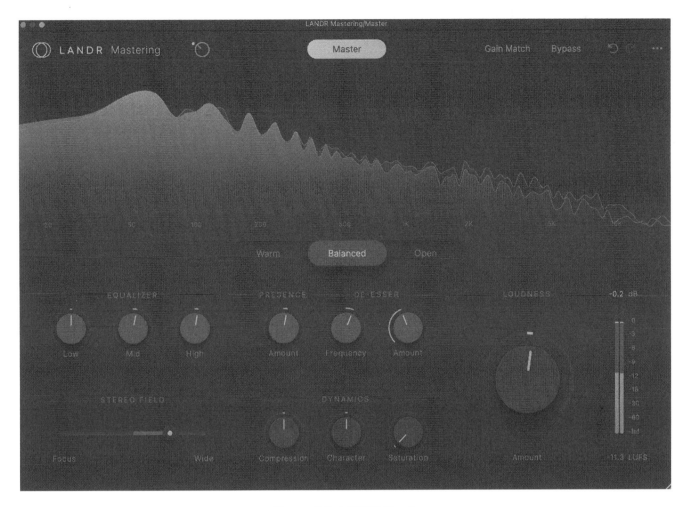

Figure 8.9: LANDR Plugin
Courtesy LANDR

WHERE TO INSERT A MASTERING PLUGIN

Any self-mastering tool will replace any and all individual compression, limiting and EQ plugins that you might already be using on your mix buss. It's best to remove these from the stereo buss and let the mastering plugin do all the work.

As a result, it doesn't matter which insert slot it occupies if it's the only plugin being used on the master buss.

If you choose to use other plugins in conjunction with your Ai mastering plugin, place the mastering plugin last in the signal chain. That way you'll be sure that there are no overloads and the output level stays exactly where it needs to be.

TAKEAWAYS

- A maximizer is a dynamics processor designed to increase the loudness of the track.

- A limiter is a dynamics processor that prevents the signal from exceeding a certain level.

- Comprehensive mastering plugins combine compression, EQ, limiting, and metering, as well as other processors, into a single plugin.

- Dedicated workflow plugins replicate the signal chains of accomplished mastering engineers or renowned studios.

- Mastering DAW plugins are extremely powerful but can do more harm than good if not used carefully.

- Additional plugins on the master buss aren't necessary if you're using a mastering plugin.

- When using an AI-based mastering plugin, rely on the AI assistant or Learn function and trust the results.

- Use a mastered reference track from the same music genre to get the best results.

9
ONLINE MASTERING SERVICES

Like mixing, mastering costs vary quite a bit. An A-list mastering engineer may charge anywhere from $250 to over $500 per hour. As a result, mastering an album at a top facility can cost anywhere from $2,000 to $10,000 or more, especially if specialized masters for vinyl, CD, or Apple Digital Masters are required.

While it's tempting to master your own music to save money, do you really want to risk your precious tracks on a process you're not confident in?

As you've read in previous chapters, your listening environment and lack of experience might prevent you from creating the high-quality final product you deserve.

That means it might be time to consider an online mastering service. Online mastering services actually fall into two categories - Pro Studio e-Mastering and Automated Online Mastering

PRO STUDIO E-MASTERING

There was a time when if you wanted your song or album mastered, you had to physically go to the studio and sit with the engineer as your project was mastered. This had its advantages—you could hear details you hadn't noticed before, and learn from the mastering engineer while you were there.

That scenario is less common today, as most pro mastering facilities now offer e-mastering, which involves electronically delivering your material to the studio for mastering.

This isn't new concept, as most high-end studios were moving files between the clients and the replication facilities since the early 2000s. Today, it's almost guaranteed that this will be the process if you choose to work with a pro mastering engineer.

That said, many top-tier facilities offer discounted rates on e-mastering in certain situations, such as:

- You don't need your project in a hurry

- You're not particular about which engineer handles your project

This can be a great deal in that you either get the main engineer (or one of the engineers in a large facility) during a slow period, or you get an assistant working under the watchful eye of the main mastering engineer.

Most studios that offer e-mastering provide multiple package options at different price points.

- The least expensive packages usually offer a master processed (equalized and compressed) by an assistant engineer or the first available mastering engineer. You'll be responsible for adapting the mastered track to the appropriate delivery format (CD, streaming, vinyl).
- A second tier usually adds a CD master, and in some cases you can choose the engineer you prefer.
- The highest tier includes everything mentioned above, plus files for streaming and Apple Digital Masters.

Where things differ between studio's offerings is the number of revisions allowed. This can go anywhere from one to unlimited, but definitely depends on the facility.

Pro studio e-mastering is a great option if you're unsure about mastering your project yourself and want a professional to handle it at an affordable rate.

AUTOMATED ONLINE MASTERING

The most affordable way to get a project mastered is by using an automated online mastering service like Landr, eMastered, or Cloudbounce.

Prices for these services range from as low as $4 per song to a monthly fee of around $100, which allows you to upload an unlimited number of songs. Most of these platforms use some form of machine learning to achieve their results.

Depending on the quality of the mix (a higher-quality mix will lead to a better final product), the results can be surprisingly good. Online mastering services typically allow you to try different mastering settings for a single price. Some platforms also let you make multiple revisions without additional charges.

Many online mastering platforms are now available, but Landr has been around the longest (see Figure 9.1). It was launched back in 2014 and has benefited from the most training, having created masters for over 5 million songs since them.

As we know, the way an AI-based platform is trained is all important and the larger the dataset it learned from, the better job it does. Landr has improved significantly over the years and can now produce results that aren't far from those of a human mastering engineer.

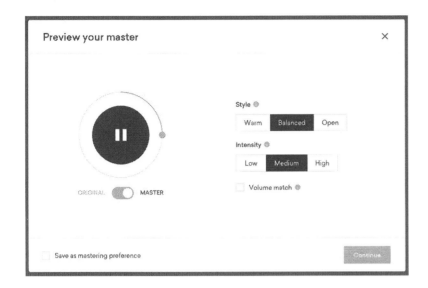

Figure 9.1: LANDR Mastering
Courtesy LANDR

There many other mastering platforms that have been released in the last few years. eMastered, Cloudbounce, Maaster, Songmastr, Soundborg, Master Channel, and Bakauge are just a few that have competed for market share in this area.

Plugin developers like Waves (Waves Mastering), Slate Digital (Virtu) and Plugin Alliance (mastering.studio) have also entered the online mastering market as well (see Figure 9.2).

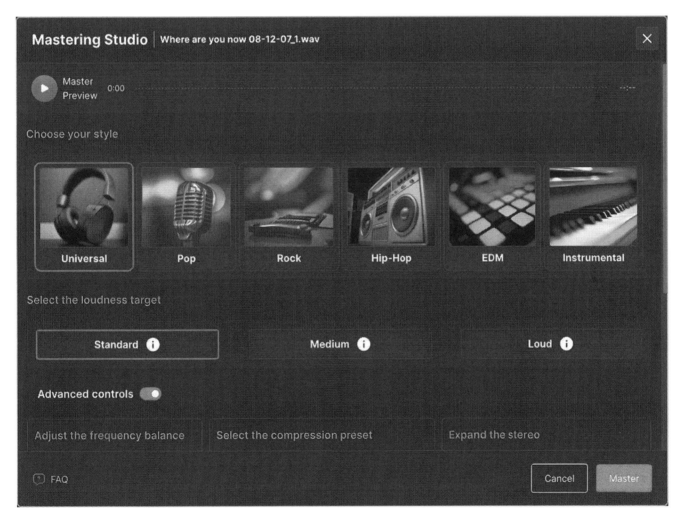

Figure 9.2: Slate Digital Virtu
Courtesy Slate Digital

111 | Online Mastering Services

Using A Reference Track

One key to creating a great master on an online mastering platform is to use a reference track. This means uploading a finished song you like to the platform, where it will analyze and replicate the EQ curve, dynamics, and loudness. It will then apply those properties to your song file and create a master that sounds similar to the reference track.

TIP: One of the secrets to selecting a good reference track is to first choose something in your genre, and secondly, make sure it's already a mastered file.

It's fine to use a major label release from a streaming service like Spotify, Apple Music, or a CD. The platform doesn't store the reference track, so you're not violating copyright—only its parameters are temporarily copied and applied to your master.

You can experiment by uploading your song to Landr, Cloudbounce, Maaster, Songmastr, or any of the popular mastering platforms, then have it create a master track without using a reference track first.

Then, upload a reference track, create another master, and compare the two. You'll likely find that the master created with the reference track sounds much better.

Limitations To Be Aware Of

While online mastering can produce surprisingly good results, there are also some caveats to be aware of.

1. **Some platforms are not equipped to master an entire album**, so you'll need to master each song individually. The downside is that this can lead to slight differences in tonal quality and level between tracks. This is one area where mastering engineers are particularly good at what they do and it's hard for a mastering AI to beat them at making a group of songs sound similar with consistent levels.

2. **Even the online mastering services that will master an album in its entirety will not creatively insert spreads (the time in between songs) and fades in between songs.** The timing of the spreads can make a difference in how the songs flow from one another, and so far there's not an online mastering platform that will do crossfades between songs. Of course, this is only important for physical products like vinyl and CDs and and doesn't apply to songs only intended for streaming.

3. **Another limitation is that many AI mastering platforms have a maximum resolution of 44.1kHz/16 bit.** While this works fine for CDs or streaming distributors, it may not meet the requirements of record labels or high-resolution platforms like Apple Music or Tidal, which

require 24-bit files with a sampling rate of 96kHz or higher. **NOTE**: *If you originally recorded your tracks at 44.1k or 48kHz, exporting to a higher sample rate won't improve their quality.*

In fact, Apple Music has a special high-resolution program called Apple Digital Masters (see Chapter 10), which requires preparation by an Apple-certified mastering engineer for submission to the platform.

4. **Yet another potential downside is if you're planning on pressing vinyl**. A mastering engineer will normally create a separate master that's not as loud and not as bass heavy to ensure that the disc cutting goes well. As of now, online mastering platforms do not have this feature, although you can experiment with alternative settings to obtain a separate master that will work for vinyl.

How To Use Online Mastering

1. Upload your song to the online mastering service
2. Select the appropriate settings for your song
3. Upload a reference track (optional, but highly recommended)
4. Select *Create Master*
5. Preview the track and make revisions if necessary
6. Download your master

List Of Online Mastering Platforms

Here are the online mastering platforms mentioned above:

- Bakauge - Bakauge.com
- Cloudbounce - Cloudbounce.com
- eMastered - eMastered.com
- LANDR - LANDR.com
- Maastr - Maastr.com
- Masterchannel - Masterchannel.ai
- Songmastr - Songmastr.com
- Mastering.studio - Mastering.studio

- Soundborg - mrmastering.com/soundborg
- Virtu - Slatedigital.com/virtu
- Waves Mastering - Waves.com/online-mastering

TAKEAWAYS

- Online mastering options include pro studio e-mastering or automated mastering platforms.
- Pro studio e-mastering offers professional-level mastering at a reduced rate, depending on the engineer and the services you require.
- Automated online mastering has improved significantly over the years due to extensive training.
- To achieve the best results, use a mastered reference track from the same genre.
- Many automated online mastering services offer a maximum resolution of just 44.1kHz/16-bit.
- Many automated online mastering services aren't equipped to master full albums.
- Even those that can master full albums can't handle nuances like custom track spreads (time between songs), crossfades, Apple Digital Masters, or vinyl masters.

10

MASTERING FOR ONLINE DISTRIBUTION

It wasn't that long ago that we really needed to know a lot about creating files for sharing online, whether directly with friends or with an online music streaming service.

Today data-compressed files like MP3s are less used than they once were, and we're quickly headed to a world where high-resolution files are the norm.

That said, it helps to know about how data compression works. Even though the process is mostly transparent to us during conversion to online streaming, there are rare occasions when we're asked to send an MP3. Of course we want it to sound as good as possible, so it pays to know the tricks to creating one.

DATA COMPRESSION EXPLAINED

Data compression is not at all like the audio compression that we've been talking about so far in this book.

Data compression is the process of reducing the size of a file by encoding, restructuring, or modifying it in some other way.

File formats like the MP3 do this in what's known as a "lossy" manner, which means that they throw away certain audio information that the encoder determines isn't important and won't be missed. Of course, if we compare a data-compressed file to its original non-data-compressed source file, we might hear a difference after than happens.

That's why the following information and parameter settings are so important—so you get a data-compressed file that sounds as close to the uncompressed source file as possible.

The encoder that performs the data compression also does the decompression as well, which is why it's known as a *codec*, which is short for compression/decompression. Table 10.1 shows a list of popular codecs currently in use online. Keep in mind that there are dozens more besides these, but most haven't risen to widespread use for music distribution.

Table 10.1: Commonly Used Audio Compression Formats

COMPRESSION	MEANS	TYPE	COMMENTS
AAC	Advanced Audio Coding	Lossy	Generally thought to be the best sounding of the lossy codecs
ALAC	Apple Lossless Audio Codec	Lossless	Supports up to 8 channels at 384kHz but mostly used for archiving
FLAC	Free Lossless Audio Codec	Lossless	The most widely used lossless codec
MP3	MPEG 2 - Audio Layer III	Lossy	Once the standard for file compression. Quality varies with different encoders.
OGG Vorbis	Derived from the jargon of the computer game Netrek	Lossy	Better sounding than MP3 but not as widely adopted. Used in many video games and by Spotify.
WMA	Windows Media Audio	Lossy and Lossless versions	PC format; not widely supported

Lossy Codecs

Lossy compression utilizes a perceptual algorithm that removes signal data that's being masked or covered up by other program data that's louder. Because this data is discarded and never retrieved, it's known as *lossy*.

This is done not only to make the audio file smaller, but also to fit more data through a small data pipe, like when an Internet service has limited bandwidth.

To illustrate what lossy compression does, think of the inner tube for a bicycle tire that's filled with air. When the air is let out, the tube takes up less space, but the same amount of rubber remains, allowing it to fit into a small box on the store shelf. This is the same idea behind lossy audio data compression.

Depending on the source material, lossy compression can be either completely inaudible or somewhat noticeable. It should be noted that even when it's audible, lossy compression still does a remarkable job of recovering the audio signal and can sound quite good with the proper preparation, as we'll soon discuss.

Lossless Codecs

The opposite of lossy compression is *lossless* compression, which never discards any data and restores it completely during decoding and playback.

FLAC is the most widely used lossless codec, although ALAC (Apple Lossless) is also commonly used.

The lossless compressed files are usually 40 to 60 percent as large as unencoded AIFF or Wav PCM files used in a DAW, but that's still quite a bit larger than the typical MP3 or AAC file, which is about a tenth of the original size.

File Compression Encoder Parameters Explained

Regardless of the encoder, there's one parameter that matters most in determining the quality of the encode, and that's *bit rate*, which is the number of bits of encoded data that are used to represent each second of audio. Lossy encoders provide a number of different options for bit rate.

Typically, the rate chosen is between 128 and 320 kilobits per second (kbps), although it could go as low as 96kbps for a mono voice-only podcast. By contrast, uncompressed stereo audio as stored on a compact disc has a bit rate of about 1400kbps (see Table 10.2).

Table 10.2: Bit rate quality comparison.

BIT RATE	QUALITY	COMMENTS
64kbps	Poor	For voice and mobile
128kbps	Poor	Minimum bit rate for music
192kbps	Good	Acceptable quality
256kbps	Very Good	Apple Music bit rate
320kbps	Excellent	Premium streaming rate

BIT RATE	QUALITY	COMMENTS
1411.2kbps (44.1kHz/16 bit)	CD Quality	Too large for streaming
2304kbps (48kHz/24 bit)	Very-High Quality	Television and movie standard
4608kbps (96kHz/24 bit)	Audiophile Quality	High-resolution audio - Apple Digital Masters suggested rate

Data-compressed files encoded with a lower bit rate will result in a smaller file and therefore will download faster, but they generally play back at a lower quality. If the bit rate is too low, compression artifacts (sounds that weren't present in the original recording) may appear during playback.

A good demonstration of compression artifacts is provided by the sound of applause, which is difficult to compress because it's so random. As a result, an encoder's limitations become more apparent, often audible as a slight ringing.

Conversely, a high bit-rate encode will almost always produce a better-sounding file, but also a larger file. This may make it too large to send via email, or take too long to download. However, with today's widespread high-speed Internet and seemingly unlimited storage, these issues are becoming less of a factor..

The norm for acceptable-quality compressed files is 160kbps. Here are the pros and cons of different bit rates used for music.

- **128kbps.** Lowest acceptable bit rate, but may have marginal quality depending on the encoder. Results in some artifacts but offers a small file size.
- **160kbps.** Lowest bit rate considered for a high-quality file.
- **320kbps.** The highest quality; larger file size, but sound can be indistinguishable from CD in certain circumstances.

Three modes are coupled with bit rate and have a bearing on the final sound quality of the encode.

- **Variable Bit Rate (VBR)** mode maintains a constant quality while raising and lowering the bit rate depending on the program's complexity. Size is less predictable than with ABR (see below), but the quality is usually better.
- **Average Bit Rate (ABR)** mode varies the bit rate around a target rate.
- **Constant Bit Rate (CBR)** mode maintains a steady bit rate regardless of the program's complexity. CBR mode usually provides the lowest-quality encode, but the file size is very predictable.

Within a given bit-rate range, VBR provides higher quality than ABR, which in turn provides higher quality than CBR.

It's All About The Source File

Encoding a wav or aiff file to create a compressed audio file may seem easy, but making it sound great requires a bit of thought, some knowledge, and a little experimentation.

The goal is to encode the smallest file with the highest quality, which, of course, is the tricky part.

TIP: The settings that might work on one particular song or type of music might not work on another.

Lossy coding makes the quality of the master mix critical because high-quality audio is less likely to be damaged by this encoding than low-quality audio. Therefore, it's vitally important that you start with the best audio quality (highest sample rate and most bits) possible.

It's also important to listen to your encoded file and try various parameter settings before settling on the final product. Listen to the encoded file, A/B it with the original, and make any additional changes you feel are necessary.

Sometimes a big, thick wall of sound encodes terribly, and you need to ease back on the audio compression and limiting of the source track. Other times, heavy audio compression can encode better than with a mix with more dynamics.

After some experience, you can make a few predictions, but you can never be certain, so listening and adjusting is the only sure way.

The Ins And Outs Of File Metadata

Some digital music files can contain not only the actual audio, but also information about that song called *metadata*. You can think of metadata as a small database associated with each song, containing tags that identify the song name, artist, album, music genre, release year, and more.

Obviously, those tags tell more about the file than a filename ever could. You could have a file called "Jjbr#$.mp3," but as long as it has accurate tags, it will be identified as "Electrolux" by SNEW from the *What's It to Ya* album.

The most common metadata fields are:

- **Title.** The track title.
- **Artist.** The artist who recorded the track.

- **Album.** Which album the track belongs to (if applicable).
- **Track.** The track number from the album (if applicable).
- **Year.** The year the track was published.
- **Genre.** The music genre—for example, speech, rock, pop.
- **Comment.** Additional notes about the track.
- **Copyright.** Copyright notice by the copyright holder.
- **Album Art.** Thumbnail of the album art or artist.

Other data that can be included in addition to these common fields are ISRC codes, web addresses, composer, conductor, and orchestra.

Metadata is supported by MP3s, Ogg Vorbis, FLAC, AAC, Windows Media Audio, and a few other less commonly used file formats.

TIP: It's more critical than ever that the metadata is accurate. It's the best way for a song to be identified when it's streamed so that the artist, label, songwriter and publisher get paid. Be sure to fill in all the metadata fields before you release your digital music file to the world. Although plenty of editors allow listeners to add data afterward, wouldn't you prefer it comes directly from you?

Creating Great-Sounding Online Files

Sometimes a small MP3 file is sufficient for showing ideas to collaborators, fans, or music supervisors, or for getting feedback on rough mixes. Even though master quality may not be a requirement, we still want it to sound as good as possible.

Here are some tips to help you get started without trying every possible parameter combination. Remember that the settings that work on one particular song or type of music might not work on another.

- **Start with the highest-quality audio file possible.** High-quality audio is less likely to be damaged during encoding than low-quality source files. Therefore, it's vitally important that you start with the best audio quality (highest sample rate and most bits) possible. That means that sometimes it's better to start with the 24-bit master or make the encode (usually an MP3) while you're exporting your mix, rather than using a 16-bit CD master as the source for your encodes.
- **Lower the level.** Files with lower levels than your CD master usually result in better-sounding encodes. A decrease by a dB or two can make a big difference in how the encode sounds.

> TIP: MP3 encoding almost always results in the encoded material being slightly louder than the original source. Limit the output of the material intended for MP3 to −1.1dB or less, instead of the commonly used −0.1 or −0.2dB, to avoid digital clipping.

- **Filter out some high frequencies.** Filter out the high frequencies at the point that sounds best (judge by ear). Most codecs (especially for MP3) have the most difficulty with high frequencies, and rolling them off frees up processing power for encoding the lower and mid frequencies. You trade some top end for better quality in the rest of the spectrum.

- **A busy mix can lose punch after encoding.** Sparse mixes, like acoustic jazz trios, seem to retain more of the original audio punch.

- **Use Variable Bit Rate (VBR) mode** when possible.

- **Turn off Mid-Side Joint Stereo, Intensity Joint Stereo,** and **Stereo Narrowing** if these parameters are available.

- **Don't use a bit rate below 160kbps** (higher is better).

- **Don't hyper-compress.** Leave some dynamic range in your mix so the encoding algorithm has dynamics to preserve.

- **Set your encoder to maximum quality for the best results.** The encoding time is negligible anyway.

> TIP: The suggestions apply to both MP3 and other online codecs, including those used for the various streaming services.

CREATING FILES FOR STREAMING SERVICES

Submitting song files to the various streaming services is usually done either through an aggregator such as TuneCore or CD Baby, or directly from a record label.

Regardless of who submits the files, the requirements are the same in most cases – 44.1kHz/16-bit audio, the same as with a CD master (although Apple Music is the exception - more on that later). In some cases a high-quality MP3 at 320kbps is also acceptable for submission, but obviously this doesn't represent the best that the music can sound.

After submission, the streaming service encodes it according to their specifications, which vary from service to service. It's good to know how the music will be encoded in order to provide the best-sounding source file, so Table 10.3 shows the current streaming specs of the most popular services.

Table 10.3: Streaming Specs for the Most Popular Services

BIT RATE	QUALITY	COMMENTS
Amazon Music	Opus FLAC FLAC	320kbps - Standard 850kbps - High Definition 3730kbps - Ultra HD
Spotify	Ogg Vorbis Ogg Vorbis AAC	96kbps - Low 160kbps - Standard quality 320kbps - Premium quality
Pandora	MP3	128kbps free 192kbps premium
YouTube	AAC	240P video - 64kbps 360P video - 128kbps 480P video - 128kbps 720P video - 192kbps 1080P video - 192kbps
Apple Music	AAC	256kbps
YouTube Music	AAC	128kbps - Normal 256kbps - High
iHeartRadio	AAC	128kbps
Deezer	MP3 MP3 MP3 FLAC	64kbps - Basic 128kbps - Standard 320kbps - High Quality 1411kbps - HiFi
SoundCloud	MP3 AAC	128kbps - Free 256kbps - Go+
Tidal	AAC MP3 FLAC	320kbps - Low 1411kbps - High Quality 3730kbps - Max

TIP: Treat files intended for streaming services as you would for an MP3 file and lower the level 1 to 2dB to ensure distortion-free encoding.

Checking The Sound Of Streaming Platforms

Although mixing at the playback level of the distributor offers no advantage, you might want to hear what your master will sound like on Spotify or Apple Music. Luckily you can now do that without resorting to posting to the platform first.

Several plugin developers now offer plugins that play back your mix at the actual platform level using the same codec. For instance, you can hear what it will sound like when posted to Spotify, Amazon Music, Apple Music, Deezer, Tidal, and most of the other major music distribution platforms.

Examples of these plugins include Nugen Audio's Mastercheck Pro, and Adaptr Audio Streamliner (see Figure 10.1).

Figure 10.1: Adaptr Audio Streamliner
Courtesy Adaptr Audio - Plugin Alliance

Be Aware Of Sound Check

In order to make sure that all music is played back at the same level, several streaming services have implemented loudness normalization. For example Apple Music uses "Sound Check" to adjust all songs to a set target gain level of -16LUFS.

Spotify uses its own normalization process that aligns everything to -14LUFS and uses a limiter when necessary to ensure peaks don't go above that, which means that the process can be more destructive to the audio. Other services have different target levels, which means that there's currently no standard.

This means a song that's heavily compressed will play back at the same level as one with lots of dynamic range. *The key difference is that the song with more dynamic range will sound better.* This is important to remember when mastering, especially when tempted to squeeze every last tenth of a dB in level.

Thanks to loudness normalization processes like Sound Check, the loudness wars are likely to end and our audio will sound better than it has in a long time.

SUBMITTING TO ONLINE STREAMING SERVICES

At one time it was possible to submit your song directly to various streaming services but that proved highly inconvenient and time consuming for both the service and the artist.

Today virtually all the major online music streaming services require a large record label account, meaning that you either need a somewhat large record label or a digital distributor to submit your songs.

TIP: Your label needs a catalog of 200 songs and a minimum of 100 songs a year to be considered for direct submission on most streaming services.

It's so much easier to use one of the many distribution services, many of which can get your songs on more than 100 services worldwide. Not only that, each online service has a different file format requirement, which can cause you to spend a lot of time with file preparation, when submission is just a single click away with CD Baby, TuneCore, DistroKid or ReverbNation (among others).

Table 10.4 gives a quick overview of a few of the more popular digital music distributors.

Keep in mind that this data was accurate as of the writing of this book, but things change rapidly in this digital world we live in so they may be complete different by the time you read this.

Table 10.4: Digital Music Distributors

	CD BABY	TUNECORE	DISTROKID	DITTO
Single Fee	$9.99	$14.99/year for unlimited releases	$22.99/year for unlimited releases	$19/year for unlimited releases
Album Fee	$9.99	$14.99/year for unlimited releases	$22.99/year for unlimited releases	$19/year for unlimited releases
Sales commission	9%	0%	0%	0% standard - 20% accelerator
Number of digital partners (Apple Music, Amazon, etc.)	150+	150+	150+	150+
Distribute music videos	No	No	Yes ($99/year)	Yes
Distribute to TikTok	Yes	Yes	Yes	Yes
Number of artists/labels	2 Million	n/a	3 Million	1 Million

In a nutshell, TuneCore, DistroKid, and Ditto Music charge annual fees but don't take a percentage of your sales. CD Baby takes a 9 percent cut but doesn't charge an annual fee. There are many other distributors and they all have similar fees and features.

> TIP: It may take as long as 2 weeks from the time you upload your song until it's released on some platforms.

Streaming Audio Requirements

As stated previously, most distributors require a CD-quality 44.1kHz/16 bit file, which means that the mastering engineer must prepare a CD master (see Chapter 12 for more info on how this is done).

In the case of high-resolution tiers offered by streamers like Apple Music, Tidal, and Amazon Music, their resolution requirements differ. Some services like Apple Music, consider any sample rate at 24 bits to be hi-res, while others set a minimum of 96kHz/24-bit (we'll get into this more later). While upsampling a 44.1kHz or 48kHz file to 96kHz results in no improvement in audio quality, it does simplify submission to the different services.

Target LUFS levels are another area of confusion. As mentioned earlier in this chapter, each streaming service has its own designated playback level. That doesn't mean that you have to submit a file at the level though! Since the streaming service will encode any file submitted to its target level using its own algorithm, *the best thing that you can do is supply a great-sounding file and ignore the LUFS level.*

> *If I master something and it sounds good at a certain volume—regardless of the LUFS target—that's where it should stay.*
>
> —Maor Appelbaum

CREATING A FLAC FILE

A format that has recently gained significant attention is the lossless FLAC format, which stands for Free Lossless Audio Codec. It works similarly to a standard MP3 file, but it's lossless like a Zip file, and specifically designed for audio.

Unlike other lossless codecs from DTS and Dolby, FLAC is non-proprietary so it's unencumbered by patents, and has open-source implementation. Additionally, FLAC has been adopted as a release format of choice by some of the world's biggest recording artists, including Pearl Jam, Nine Inch Nails, the Eagles, and even reissues from The Beatles.

FLAC supports a bit depths from 4 to 32, and up to 8 channels. Even though it can support any sampling rate from 1Hz to 655,350Hz, you don't need to specify a bit rate because it automatically determines it from the source file.

Plus it has a "cue sheet" metadata block for storing CD-style tables of contents and track and index points. It's an excellent way to deliver high-fidelity music files with relatively small file sizes, though it's not yet supported by all applications or players.

Although many DAWs don't have a FLAC encoder built in, there are a number of players and encoders that can be downloaded for free, as well as QuickTime playback components and Apple Music scripts.

APPLE DIGITAL MASTERS

With Apple Music now such a huge part of online music distribution, knowing the latest on how it treats your audio can be beneficial to the final audio quality of your file. Apple Music actually has two different categories: standard Apple Music audio and Apple Digital Masters high-resolution audio. Let's look at both.

A Look At AAC, The Apple Music File Format

Apple Music uses the AAC file format (which stands for Advanced Audio Coding) as a standard distribution format. Contrary to popular belief, it's not a proprietary Apple format. In fact, it's part of the MP4 specification and generally delivers excellent-quality files that are about 30 percent smaller than a standard MP3 of the same data rate.

All new music destined for Apple Music is encoded at 256kbps at a Constant Bit Rate with a 44.1kHz sample rate, with the exception of Apple Digital Masters, which we'll discuss below.

Apple Music Sound Check

As mentioned earlier, Apple Music uses a feature called Sound Check, which scans your library and normalizes volume so all songs play at the same level.

When you encode a song using Apple Music, the Sound Check level is stored in the song's ID3 tags. Sound Check is designed to work with MP3, AAC, WAV, and AIFF files and won't work with a file that Apple Music can't play. The audio data of the song is never changed.

Sound Check always defaults to On on all Apple devices. This does cause a dilemma for artists, producers, record labels, and mastering engineers though.

Like on other services, Sound Check will automatically lower a very loud song and boost a quiet one so they play at the same level. *Because the loud track has fewer dynamics, it can sound lifeless and even quieter than the less-compressed track due to fewer peaks.*

This has led people to rethink the value of a loud master, as it may be counterproductive in a level-controlled environment.

Spotify is another service that uses a form of Sound Check, and other services use something similar as well.

TIP: Keep Sound Check in mind when mastering for online distribution. Lowering the level by a few dB can result in a better-sounding and even louder master in this environment.

The Apple Digital Masters Format

Apple Digital Masters is an alternative to the standard Apple Music quality, where it accepts high-resolution master files and provides higher-quality AAC encodes as a result.

Music files that are supplied at 96kHz/24-bit will have an Apple Digital Masters icon placed beside them to identify them as such, although any sample rate that's a 24-bit file is also considered high-resolution (see Figure 10.2).

Figure 10.2: Apple Digital Masters logo
Source: Apple Inc.

Apple Digital Masters (formerly called Mastered For iTunes or MFiT, for short) doesn't necessarily mean that the mixer, producer, or mastering facility does anything special to the master except to check what it will sound like before it's submitted to Apple Music.

However, not just any engineer or mixer can do this. To qualify for Apple Digital Masters submission, the master must be created by a certified mastering studio. Most digital music distributors have the ability to submit to the Apple Digital Masters program as well.

Apple handles all encoding for Apple Digital Masters, not the mastering engineer, record labels, or artists.

According to Apple, this ensures consistency and prevents anyone from gaming the system by hacking the encoder. It also avoids potential legal issues when a mixer, producer, or mastering house sends files directly to Apple Music without the label's permission or uses different submission specs.

Apple Digital Masters is only an indication that a high-res master was supplied; it's not a separate product. There will always be only one version of the song on Apple Music, and it will be available at the same price regardless of whether it's an Apple Digital Masters or a normal submission.

Apple Digital Masters doesn't allow you to charge more, nor does Apple Music charge extra. Everything remains the same; you simply provide a high-resolution master for better sound.

The Apple Digital Masters Tools Package

Although the mastering engineer doesn't encode directly, Apple provides tools to hear what the final product will sound like once encoded. That way, any adjustments can be made to the master before it's submitted to Apple Music to ensure that the encode will provide the highest quality.

You can find these tools, as well as a PDF explaining mastering for Apple Digital Masters in depth, at apple.com/itunes/apple-digital-masters.

The Apple Digital Masters Tools include AURoundTrip, Mastering Droplet, afconvert, afclip, and Audio To WAVE Droplet. Except for afclip, these tools are for listening to what a song might sound like on

Apple Music. When they were first released back in 2012 they were very useful, but now there are better tools that are much easier to use.

The afclip Tool

Of all the tools that Apple supplies, probably the most important one for the mastering engineer is afclip, as it's a great check to be sure that there are no digital overs in your song submission. It works by examining an audio file and identifying areas where clipping has occurred.

It accepts audio files as input and outputs a stereo sound file containing the left channel of the original file and a right channel with graphically represented impulses corresponding to each clipped sample in the original. This sound file can then be loaded into a DAW so that you can see a visual map to locate any clipping that may have occurred.

Apple Music won't reject a master file based on the number of clips the file contains. This tool is there just so you can make an informed decision about whether to submit an audio file or go back to the drawing board and make adjustments, which is a creative decision that's entirely up to you.

TIP: afclip will find many instances of clipping that are inaudible. Unless you're hearing clipping that you're trying to track down, using this app can be a lesson in futility, especially with a client who just wants the loudest product in spite of any clipping.

The fact is that mastering engineers don't use these tools much. There are easier ways to hear what your song will sound like on Apple Music, and we now have excellent metering tools that provide the same information as afclip without the hassle of using the command-line Terminal app.

Submitting To Apple Music

It's not possible to submit directly to Apple Music if you're an indie artist, a band, a producer, or a small label. Apple reserves that feature for larger labels with hundreds of titles in their catalog that release titles regularly.

However, you can easily get your music on Apple Music by using a digital distributor, such as TuneCore, CD Baby, DistroKid, ReverbNation, Ditto, or others.

To submit directly to Apple Music, labels must pass the requirements for submission, then they use a program known as iTunes Producer or Apple's Transporter app to submit individual songs and albums. These free programs not only allow uploading product to Apple Music, but also allows the label to input all the metadata associated with the project.

This is one of the reasons why mastering engineers can't directly submit songs to Apple Music on behalf of their clients. The use of the program is exclusive to the label for their product only, and any payments from sales will only go to the bank account associated with that particular Apple Music Producer account.

> ***TIP: To display the Apple Digital Masters badge on Apple Music, you must supply the highest-resolution AIFF or WAV file, and it must come from an Apple-certified mastering engineer.***

OTHER HIGH-RESOLUTION PLATFORMS

Even though Apple Music was the first to delve into high-resolution audio back in 2012, it isn't the only platform with high-resolution streaming options. Tidal, Deezer, and Amazon Music also have high-resolution streaming available.

In fact, Amazon Music has two hi-res tiers; High Definition and Ultra High Definition. The HD tier is for CD quality 44.1kHz/16 bit audio, while Ultra HD goes all the way up to 192kHz/24 bit.

That said, there are other high-resolution platforms that offer streaming, but primarily concentrate on sales of files. Qobuz, HighResaudio.com, and HD Tracks all offer large catalogs of 24 bit songs for sale. HD Tracks offers tracks with up to 352kHz/24 bit resolution.

Other sites that offer high-resolution audio either as DSD, FLAC or ALAC files include Acoustic Sounds SuperHiRez, ProStudioMasters, and Native DSD and Beyond.

Direct Stream Digital (DSD)

When the Super-Audio CD was first introduced in 1999, it was heralded as a new age in audio reproduction, mostly thanks to the encoding process known as Direct Stream Digital (DSD).

Despite the massive marketing efforts by its creators Sony and Philips (who also created the CD), SACD never achieved the market penetration that was expected, and today it has gone the way of CDs, as fewer consumers want to invest in a shiny plastic disc. That being said, DSD files are still desirable, at least among many audiophiles, and can be downloaded from a number of online sites.

DSD differs from other analog-to-digital processes by measuring whether a waveform is rising or falling, rather than measuring the waveform at discrete points in time like in PCM (see Figure 10.3).

In current systems, this one bit is then decimated into LPCM, causing a varying amount (depending on the system) of unwanted audio side effects (such as quantization errors and ringing from the required brick-wall filter). DSD simplifies the recording chain by recording the one bit directly, thereby reducing the unwanted side effects. Look at DSD-Guide.com for more information.

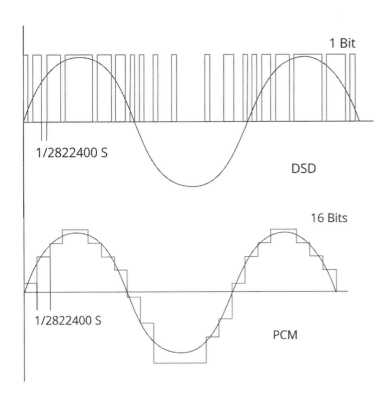

Figure 10.3: PCM versus DSD
©2024 Bobby Owsinski

Indeed, on paper DSD looks impressive. A sampling rate of 2.8224MHz (which is 64 times 44.1kHz, in case you're wondering) yields a frequency response from DC to 100kHz with a dynamic range of 120dB. Most of the quantization errors are moved out of the audio bandwidth, and the brick-wall filter, which haunts current LPCM systems, is removed.

To enable a full 74 minutes of multi-channel recording, Philips also developed a lossless coding method called *Direct Stream Transfer* that provides a 50 percent data reduction. The original DSD sampling rate has since given way to a higher rate of 5.6448MHz, which is now offered by some recorders, including those in the Korg range of products.

Even though many agree that DSD sounds superior to the LPCM technology used in most of the audio world today, there are severe limiting factors when it comes to using it. First of all, processing and editing DSD natively in a workstation isn't easy.

There are only two DAWs being made today, Merging Technologies' Pyramix and the Sonoma (which was developed by Sony). Other systems (such as Korg's AudioGate) transcode the DSD stream to LPCM at 96/24 or 192/24 for editing and processing, then back to DSD again.

One of the downsides of the format is that separate DSD D/A convertors are required. As a result, these systems can be quite a bit more expensive than a normal PCM mastering system, and considering the limited market, usually aren't worth the investment for most mastering engineers. That said, many "golden ears" claim that it's the next best thing to analog.

TAKEAWAYS

- Data compression means that the file size is reduced by encoding, restructuring, or modifying it.

- Lossy compression utilizes a perceptual algorithm that removes signal data that's being masked by louder program data.

- Lossless compression retains all data and fully recovers it during decoding and playback.

- The higher the bit rate, the better the audio quality, but it also results in a larger file.

- Metadata is a small database linked to each song, identifying the song name, artist, album, genre, release year, and more.

- Full metadata submission is essential for proper credits and for the artist, songwriter, label, and publisher to get paid.

- The majority of streaming services do not allow song submission directly, so a distributor must be used.

- Most online streaming services require 44.1kHz/16 bit CD masters for submission.

- There is no need to master to a target LUFS level as the streaming service will re-encode it anyway.

- Apple Digital Masters is an alternative to the standard Apple Music quality, where it accepts high-resolution master files (up to 96kHz/24 bit) and provides higher-quality AAC encodes as a result.

- There are a number of other streaming platforms besides Apple Music that supply high-resolution audio.

- There are a number of sites online where you can purchase and download high-resolution audio files with resolution as high as 384kHz/24 bit.

- Direct Stream Digital (DSD) differs from other analog-to-digital processes by measuring whether a waveform is rising or falling, instead of measuring it at discrete points in time. There are sites online that specialize in this format.

11

MASTERING FOR VINYL

Although it seems like almost an ancient technology in these days of 1's and 0's, the vinyl record is making a resurgence in the marketplace, increasing in sales dramatically. With annual sales of around 9 million units a year or so (the ones that are counted, at least), vinyl is no threat to other music distribution formats and probably never will be again. Still, the format is in no danger of dying either, and there are still many requests for vinyl masters every year.

While the vast majority of mixing and mastering engineers won't be purchasing the gear to cut vinyl anytime soon, it's still important to know what makes the format tick in order to get the best performance if you decide to make records to complement a CD or online project.

Before we get into the mastering requirements for vinyl, let's take a look at the system itself and the physics required to make a record. While this is by no means a complete description of the entire process of cutting a vinyl record, it is a pretty good overview.

A BRIEF HISTORY OF VINYL

It's important to look at the history of the record because in some ways it represents the history of mastering itself. Until 1948, all records were made of shellac, were 10 inches in diameter, and played at 78 revolutions per minute (RPM).

When Columbia Records introduced the vinyl 12 inch 33⅓ RPM disc in 1948, the age of hi-fidelity actually began since the sonic quality took a quantum leap over the previous generation. However, records of that time had a severe limitation in that they only held about 10 minutes of playing time per side, as the grooves were relatively wide to fit all the lower audio frequencies.

To overcome this time limitation, two refinements were made. First, in 1953 the Recording Industry Association of America (RIAA) instituted an equalization curve that narrowed the grooves, thereby allowing more of them to be cut on the record, which increased the playing time and decreased the noise at the same time.

This was achieved by boosting the high frequencies by about 17dB at 15kHz and cutting the lows by 17dB at 50Hz during the cutting process (see Figure 11.1). The opposite curve is then applied during playback. This is what's known as the RIAA curve. It's also why it sounds so bad when you plug your turntable directly into a mic or line input of a console without a phono preamp in between.

Without the RIAA curve applied, the resulting sound is thin and tinny due to the overemphasized high frequencies and attenuated low frequencies.

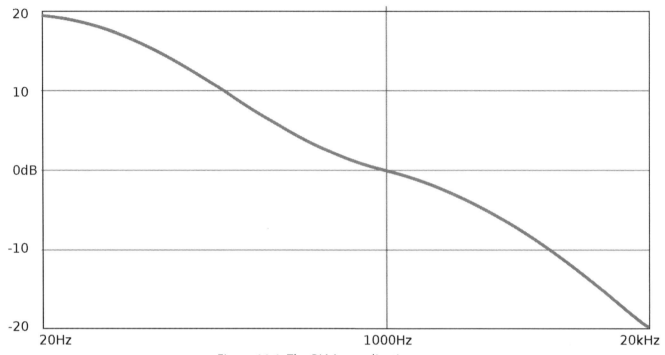

Figure 11.1: The RIAA equalization curve

The second refinement was the implementation of variable pitch, which allowed the mastering engineer to change the number of grooves per inch according to the program material. In cutting parlance, pitch refers to the rate at which the cutter head and stylus travel across the disc.

By varying this velocity, the spacing between grooves can also be changed. These two advances increased the playing time to the current 25 minutes or so (more on this later) per side.

In 1957, the stereo record became commercially available and really pushed the industry to the sonic heights that it has reached today.

HOW A VINYL RECORD WORKS

To understand how a record works, it's essential to understand what happens within a groove. If we were to cut a mono 1kHz tone, the cutting stylus would swing side to side in the groove 1,000 times per second (see groove pictures 11.2 through 11.15). The louder the signal, the deeper the groove needs to be cut.

While this works well in mono, it doesn't work for stereo, which was a major problem for many years. As mentioned earlier, stereo records were introduced in 1957, but the stereo record-cutting technique was actually proposed in 1931 by famed audio scientist Alan Blumlein.

His technique, called the 45/45 system, was revisited 25 years later by the Westrex Corporation (the leader in record-equipment manufacturing at the time) and led to the eventual introduction of the stereo disc.

Essentially, a stereo disc combines the side-to-side (lateral) motion of the stylus with an up-and-down (vertical) motion. The 45/45 system rotated the axis 45 degrees relative to the plane of the cut.

This method actually has several advantages. First, mono and stereo discs and players become compatible, and second, the rumble (low-frequency noise from the turntable) is decreased by 3dB.

Let's take a look at a few groove pictures provided by Clete Baker of Studio B in Omaha, Nebraska to illustrate how the system works.

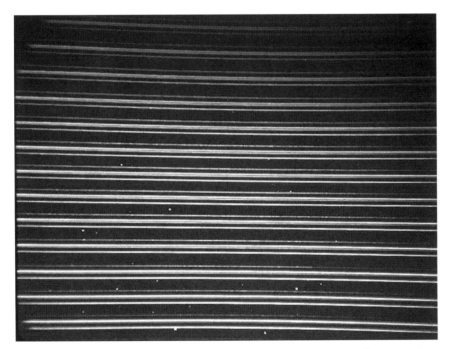

Figure 11.2: This is a silent groove with no audio information. The groove width across the top of the "v" from land to land is 0.002 inches, and the groove depth is approximately the same as the width for this particular stylus (made by Capps).
Courtesy of Clete Baker

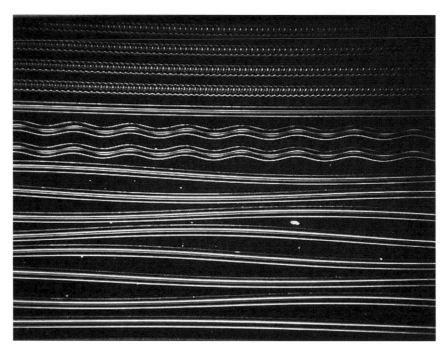

Figure 11.3: From the outside of the disk (right-hand side) going inward, you can see a low-frequency sine wave, a mid-frequency sine wave, and a high-frequency sine wave, all in mono (lateral excursion). All frequencies are at the same level at the head end of the system, which is prior to application of the RIAA curve. This demon strates that for any given level, a lower frequency will create a greater excursion than a high frequency, and as a result will require greater pitch to avoid intercuts between grooves.
Courtesy of Clete Baker

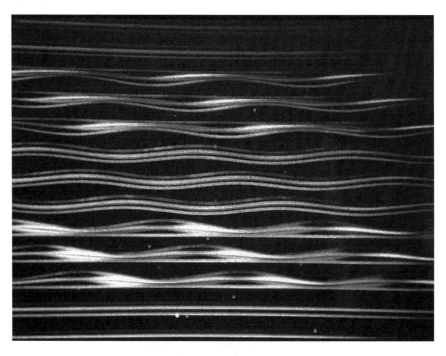

Figure 11.4: This is a sine wave applied to the left channel only toward the outer part of the record, summed to mono in the center of the view, and applied to the right channel only toward the inner part of the record. You can easily see the difference between the purely lateral modulation of the mono signal and the vertical of the left- and right-channel signals.
Courtesy of Clete Baker

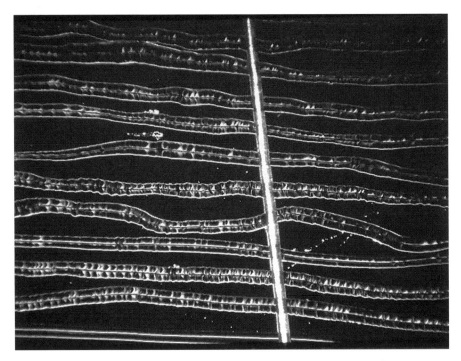

Figure 11.5: A human hair laid across the groove offers a point of comparison.
Courtesy of Clete Baker

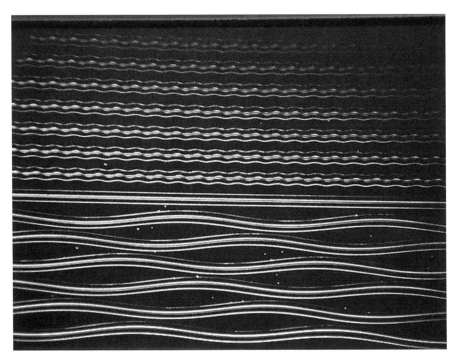

Figure 11.6: Again, lower-frequency and higher-frequency sine waves demonstrate that more area of the disk is required to accommodate the excursion of lows than of highs.
Courtesy of Clete Baker

Figure 11.7: This figure shows variable pitch in action on program audio. To accommodate the low-frequency excursions without wasting vast amounts of disk real estate, variable pitch is employed to spread the groove in anticipation of large excursions. This narrows the groove if the material doesn't contain any low frequencies that require it.
Courtesy of Clete Baker

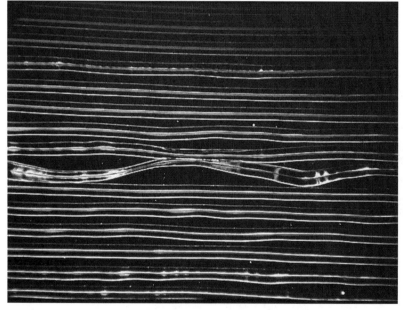

Figure 11.8: This is what happens when variable pitch goes bad, which is a lateral intercut caused by insufficient variable pitch for a wide lateral excursion. Toward the bottom center of the slide, the outside wall of the loud low frequency has touched the adjacent wall of the previous revolution, but the wall has not broken down, and a safe margin still exists so it won't cause a skip. However, on the next revolution an excursion toward the outside of the disk has all but overwritten its earlier neighbor, which is certain to cause mistracking of the playback stylus later.
Courtesy of Clete Baker

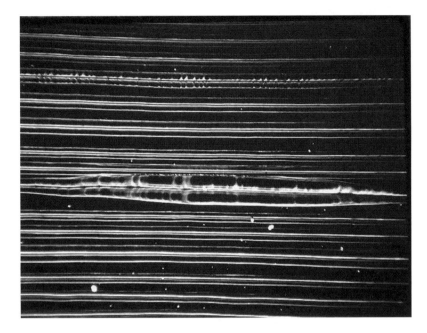

Figure 11.9: Lateral excursions aren't the only source of intercuts. This shows a large low-frequency vertical excursion caused by out-of-phase information, which is not severe enough to cause mistracking, but will probably cause distortion.
Courtesy of Clete Baker

TIP: This is exactly why low frequencies are panned to the center, or an elliptical equalizer (see later in the chapter) is used to send them to the center, during disc mastering.

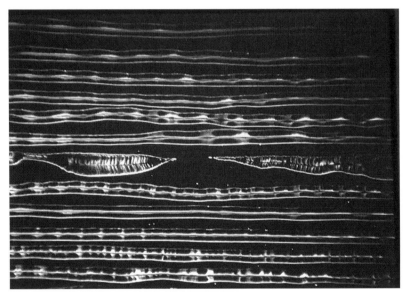

Figure 11.10: Large vertical excursions can cause problems not only by carving out deep and wide grooves that result in intercuts, but by causing the cutting stylus to literally lift right off the disk surface for the other half of the waveform. This would cause a record to skip.
Courtesy of Clete Baker

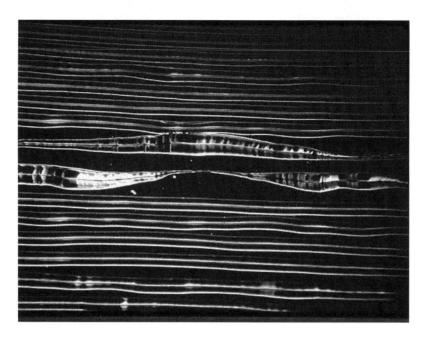

Figure 11.11: Here a near lift is accompanied on the following revolutions by lateral intercut, which will result in audible distortion. To solve this problem, the mastering engineer can increase groove pitch and/or depth, lower the overall level at which the record is cut, reduce the low-frequency information, sum the low frequencies at a higher crossover point, or add external processing, such as a peak limiter. Each of these can be used alone or in combination to cut the lacquer, but none can be employed without exacting a price to the sound quality.
Courtesy of Clete Baker

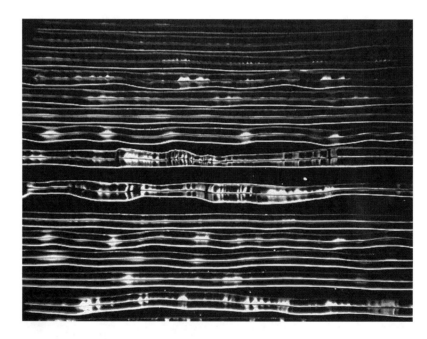

Figure 11.12: Here is the same audio viewed in Figure 8.11 only after processing. In this case a limiter was employed to reduce dynamic range (the surrounding material is noticeably louder as well) and rein in the peaks that were causing intercuts and lifts. This section is cut more deeply in order to give vertical excursions plenty of breathing room. Pitch too has had to be increased overall to accommodate the slightly wider groove, despite the reduced need for dramatic dynamic increases in pitch due to the reduction of peaks by the limiter.
Courtesy of Clete Baker

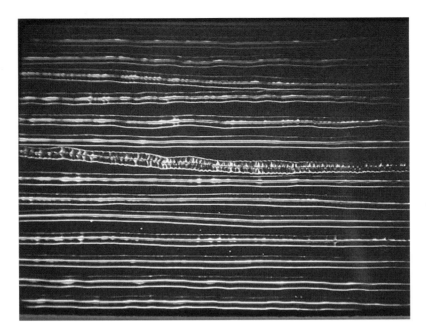

Figure 11.13: Because of the physics of a round disk, high-frequency information suffers terribly as the groove winds closer to the inner diameter. Here what high-frequency-rich program material near the outer diameter of the disk looks like.
Courtesy of Clete Baker

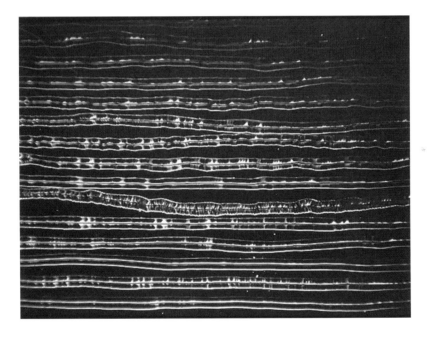

Figure 11.14: Here is the same audio information as in Figure 8.13 only nearer the inside diameter of the disk.
Courtesy of Clete Baker

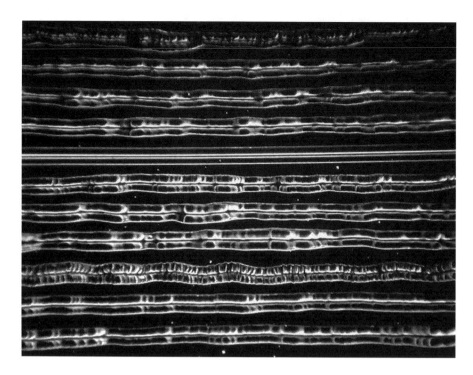

Figure 11.15: The ideal: normal, healthy-looking program audio.
Courtesy of Clete Baker

THE VINYL SIGNAL CHAIN

While the signal chain for vinyl is similar to that of a CD, there are some important distinctions and unique components involved. Let's look at the chain from the master lacquer (the record that is cut to send to the pressing plant) on back.

The Master Lacquer

The master lacquer is the record that is cut to send to the pressing plant. It consists of a mirror-smooth aluminum substrate coated with cellulose nitrate (a distant cousin to nitroglycerine), along with resins and pigments to keep it soft and aid in visual inspection (see Figure 11.16).

The lacquer is extremely soft compared to the finished record, and the master can never be played once it's cut. To audition the mastering job before cutting a lacquer, a reference disc, called a 'ref' or an acetate, is created. Since it's made of the same soft material as the master lacquer, it can only be played five or six times before the quality degrades significantly.

A separate lacquer master is created for each side of the record. The lacquer is always larger than the final record (a 12-inch record requires a 14-inch lacquer), so repeated handling doesn't damage the grooves.

Figure 11.16: A master lacquer
©2024 Bobby Owsinski

The cutting stylus, which is made of sapphire, sits inside the cutter head, which contains several large drive coils (see Figure 11.17). The drive coils are powered by a set of very high-powered (typically 1,000 to 3,500 watts or higher) amplifiers. The cutting stylus is to enable an easier and quieter cut.

The Lathe

The lathe includes a precision turntable, the carriage holding the cutter head assembly, and a microscope to inspect the grooves and make adjustments that determine their number and depth.

Although lathes by Scully and Neumann haven't been made since the early '90s, they remain highly desirable (see Figure 11.17), though new models by Agnew Analog and Sillitoe Audio Technology are now in production.

Figure 11.17: Neumann VMS-80 with SX 84 cutter head from 1984

Way back in the '50s, the first cutting systems weren't very powerful. They only had maybe 10 or 12 watts of power. Then, in the '60s, Neumann developed a system that brought it up to about 75 watts per channel, which was considered pretty cool at the time. In the '70s, the high-powered cutting systems came into being, which were about 500 watts. That was pretty much it for a while, since it made no sense to go beyond that because the cutter heads really weren't designed to handle that kind of power anyway. Even the last cutting system that came off the line in about 1990 at Neumann in Berlin hadn't really changed other than it had newer panels and prettier electronics.

—David Cheppa of Better Quality Sound

In the physical world of analog sound, all the energy is in the low end. In disc cutting, however, the opposite is true, with all the energy concentrated in the upper audio spectrum. As a result, everything from about 5kHz up begins to require a great amount of power to cut, and this is why disk-cutting systems need to be so powerful.

The problem is that if something goes wrong at that power level, it can be devastating, with the lacquer, stylus, or even the cutter head at risk of being destroyed.

The Mastering Console

The mastering console for a disc cutting system is equal to that used today for mastering in sound quality and short signal path, but that's where the similarity ends. Due to the unique requirements of cutting a disc and the manual nature of the task (as computerized gear was unavailable at the time), several features found in disc cutting have fallen by the wayside in the modern era of mastering (Figure 11.18).

Figure 11.18: A Neumann SP-75 vinyl mastering console

The Preview System

Chief among them is the preview system, an additional monitoring path made necessary by the volatile nature of disc cutting. Here's the problem—disk-cutting is essentially a non-stop operation. Once you begin cutting you must make all your changes on the fly without stopping until the end of the side.

If a portion of the program contains excessive bass, a loud peak, or something out of phase, the cutter head could cut right through the lacquer to the aluminum substrate. This would destroy not only the lacquer, but maybe an expensive stylus as well. Hence the need for the mastering engineer to hear the problem and make the necessary adjustments before any harm comes to the disk.

Enter the preview system. Essentially, the program going to the disc is delayed. Before digital delays existed, an ingenious dedicated mastering tape machine with two separate head stacks (program and preview) and an extended tape path (see Figure 11.19) was used. This gave the mastering engineer enough time to make the necessary adjustments before any damage was done to the disk or system.

Figure 11.19: A Studer A80 tape machine with preview head

Equalization

Since a disc had to be cut on the fly and computer automation was still years away, a system had to be created in order to make EQ and compression adjustments from song to song quickly, easily, and, most necessarily, manually. This was accomplished by having two of each processor with their controls stepped so that adjustments could be repeatable.

The mastering engineer would then run down all the songs for one side of the LP and mark down the EQ settings required for each song on a template. Then, as the first song was being cut through the "A" equalizer, they would preset the "B" equalizer for the next song. As song 2 was playing through the B equalizer, they would preset equalizer A for song 3, and so on (Figure 11.18).

Although this method was crude, it was effective. Naturally, today it's much easier, because the master is a file in a workstation where the EQ and compression have already been added.

The Elliptical Equalizer

One of the more interesting relics from the record-cutting days is the elliptical equalizer, or low-frequency crossover. This unit redirects all low frequencies below a preset frequency (usually 250, 150, 70, or 30Hz) to the center.

This prevents excessive lateral movement of the cutting stylus due to low-frequency energy on one side or out-of-phase material. Obviously, this device could negatively affect the sound of a record, so it had to be used with care.

HOW RECORDS ARE PRESSED

Pressing records is such a primitive process by today's standards that it's pretty amazing that they sound as good as they do. It's a multi-step process that's entirely mechanical and manual, with many areas that can negatively influence the final product. Here are the steps from the master lacquer to the final pressed record.

1. **The master lacquer** is used as the first of several metal molds for pressing vinyl records. The lacquer is first coated with a layer of tin and silver nitrate and then placed in a nickel sulfamate bath and electroplated. The lacquer is removed from the bath, and the nickel coating is peeled away. The separated nickel is what's known as the *metal master* and is a negative of the lacquer.

2. **The metal master** is placed back into the nickel bath and electroplated a second time. The separated metal part, known as the mother, is a positive copy with grooves that can technically be played (though doing so would destroy it).

3. **The mother** is then placed back into the nickel bath for yet another electroplating. The resulting separated metal part, known as the **stamper** (see Figure 11.20), is a negative copy bolted into the record presser to stamp out the vinyl records.

   It's worth noting that, just like tape, each successive copy is a generation down, reducing the signal-to-noise ratio by 6dB. Also, great care must be used when peeling off the electroplating, since any leftover material will result in a pop or click on the finished product.

Figure 11.20: A metal stamper

147 | Mastering For Vinyl

4. **The vinyl** used to make records actually comes in a granulated form called *vinylite,* and it's naturally honey-colored, not black. Before pressing, it's heated into a modeling clay-like form and colored with pigment. At this point it's known as a *biscuit*.

5. **The biscuit** is placed in a press, resembling a large waffle iron, and heated to around 300 degrees. Temperature is important because if the press is too hot, then the record will warp; if it's too cold, the noise levels will increase. After pressing, excess vinyl is trimmed with a hot knife, and the records are placed on a spindle to cool at room temperature.

All these metal parts wear out, with a stamper becoming dull after around 70,000 pressings. As a result, several sets of metal parts must be made for large orders, and for the best-selling records of the past, even multiple lacquers.

For some great lathe pictures, visit aardvarkmastering.com/history.htm, and for a full visual explanation of the record pressing process go to https://www.furnacemfg.com/how-to-make-vinyl-record.

> *The beauty of vinyl is that artists, producers, and clients come in with a vision, even if they have zero technical knowledge. Some don't even own a turntable, but they come in with a clear idea of how they want their songs ordered and how the album should be presented. Vinyl kind of demands that level of thought. It makes them focus on creating an experience for whoever's going to be listening, and I think that's the biggest shift we're seeing.*
>
> —Eric Boulanger

NEW ADVANCES IN VINYL TECHNOLOGY

The technology of record pressing was virtually stagnant for over 40 years, with lathes and record pressing machines relying on cannibalized parts from old worn-out equipment to keep them going.

Thanks to the resurgence in vinyl sales, which injected more money into this side of the business, brand new and improved equipment is being manufactured, along with some major technological advancements. Let's take a look at some of these new developments.

New Record Presses

Until recently, if a pressing plant wanted to add another record press, they had to scour the world for forgotten pressers in old warehouses. While this did happen occasionally, most stampers found this way weren't in good enough condition to be used immediately, as they required extensive refurbishing, which took considerable time and expertise.

For the first time, several companies are now releasing brand new record stamping machines, which pressing plants are eagerly acquiring (the backlog at most facilities is about three months).

Toronto-based Viryl Technologies' Warm Tone press is computerized and fully automated, allowing a record to be stamped every 25 seconds—about half the time of a traditional press. This is something that the vinyl record community has longed for and it has finally come to pass.

Ecology-Minded Pressing

As you've seen, making a vinyl record is a messy, time consuming process. It involves toxic chemical baths, large mechanical presses, stampers that wear out easily, and perhaps worst of all, the final product is made from a petroleum byproduct.

Record pressing has shown small improvements over the years, but other than Viryl Technologies Warm Tone mentioned above, it's still done the way it was 50+ years ago.

That said, new injection moulding process invented by the German company Sonopress called EcoRecord, promises not only to cut production costs, but to improve the sound quality, and reduce the environmental impact of conventional record pressing as well.

In a conventional press, a PVC puck is heated with steam until soft, then placed between two stampers and pressed for about eight seconds. It takes about 20 seconds to press, followed by another 16 seconds for the record to cool before the process can begin again.

In the new process, a recyclable PET mixture is pre-heated, injected between two stampers, then pressed for a few seconds and cooled for another 20 seconds to ensure the mixture reaches the outer edges of the stampers.

There are several big advantages with injection moulding. First of all, the amount of energy used is cut by up to 85%. There's no excess vinyl to trim, and the stampers last much longer before degrading.

Currently, a stamper may last for as few as 2,000 records before needing replacement (although the figure is usually higher). Yet another happy byproduct is that the record's surface noise is reduced by up to 10dB over conventionally pressed records.

This seems like a slam dunk, but there are still a few challenges to overcome. So far, injection moulded records appear to be less durable, showing signs of wear after 35 plays compared to 50 for a traditional vinyl record. It can also be somewhat more expensive.

That's not the only new system with ecological promise. British Evolution Music has introduced the world's first bio-plastic record, which uses natural sugars and starches instead of plastic. Another British startup, Elastic Stage, is also working on this technology.

TIPS FOR ORDERING VINYL RECORDS

There was a point in time in 2021 and 2022 where pressing plants were so busy that it would take nine months to fulfill an order. Today that's no longer the case, as new major and local pressing plants have sprung up around the world.

Shorter delivery times are a great incentive to order vinyl, but there are a few things to keep in mind before you make that order.

1. Order directly from a pressing plant rather than through a broker. That way you have direct communication if there's a problem, and your prices will be cheaper as well.

2. Remember that you'll need a master for each side of the record that must be sequenced with the correct song order and contain the spreads (time between each song).

3. Whenever possible, place the quietest songs at the end of each side.

4. Vinyl records have a finite duration where the audio quality is at its best, which is around 23 minutes per side. For every minute beyond that, the volume will decrease, and the disc will become noisier with reduced low end, particularly toward the end of the side. In order to keep your timing within limits you can:

   - Remove a song or two to create shorter and more balanced sides
   - Edit a few songs to make them shorter
   - Consider releasing a two record set with the original song lengths. This can even be a three-sided set, which will be cheaper than the typical four sides.
   - Live with the noise and distortion that comes with sides longer than 23 minutes.

5. Colored vinyl may look great, but it can sound slightly worse than traditional black vinyl because it requires a higher extrusion temperature.

The purpose of this chapter is to illustrate how mechanical and intense the process of cutting and stamping a vinyl record can be.

Against all odds, vinyl now seems poised to stay with us for some time, so chances are you'll encounter it in the future. The more you understand the process, the more likely you are to achieve the best possible product.

TAKEAWAYS

- The Recording Industry Association of America (RIAA) equalization curve provides narrower grooves, allowing more to be cut on the record. This increases playing time while reducing noise.

- An elliptical equalizer directs low-frequency information to both grooves, enabling a louder record with improved low-end response.

- The master lacquer is the record cut by the mastering engineer and sent to the pressing plant for replication. A separate lacquer is required for each side.

- To audition the mastering job before cutting a lacquer, a reference disc, known as a 'ref' or an acetate, is created.

- Since cutting a lacquer is a continuous operation, a preview system is used to give the mastering engineer enough time to make adjustments before any damage occurs to the disc or system.

- Creating a vinyl record is a multi-step, entirely mechanical and manual process, requiring a master lacquer, metal master, mother, and stamper to produce the final record.

- The vinyl used to make records comes in a granulated form called vinylite, and it's naturally honey-colored, not black.

- The process of making vinyl records involves toxic chemical baths, large mechanical presses, and stampers that wear out easily. The final product is made from a petroleum byproduct.

- Several companies are now producing more eco-friendly records that use less heat, fewer chemicals, and, in some cases, even bio-friendly plastics.

- Whenever possible, place the quietest songs at the end of each side.

- Vinyl records have a finite period of optimal audio quality, typically around 23 minutes per side.

12

MASTERING FOR CD

For a number of years, CD sales and interest among music consumers have been declining, but recently there's been a slight upsurge again. While CDs might never return to their former sales levels, who's to say there won't be a resurgence, much like vinyl, sometime in the future? That's why it's important to have at least a basic understanding of the format and keep a reference handy if needed.

Mastering for CD requires the engineer to know far more than the basics of EQ, dynamics and editing. In fact, a proper and efficient job entails awareness of many additional processes, from inserting start/stop and track identification codes, to making choices for the master delivery medium, to checking the master for errors. In this chapter we'll look at all those things and more that are involved in modern CD mastering.

CD BASICS

The Compact Disc (the proper name for the CD) was developed in the mid-'70s as a joint venture between Dutch technology giant Philips and Japanese tech powerhouse Sony, which established the standard for both the disc and players.

The first commercial CD, ABBA's *The Visitors*, was released in 1982, although Billy Joel's *52nd Street* received the first worldwide release. More than 200 billion have been sold since then.

CDs have a digital audio resolution of 44.1kHz and 16 bits. While some of today's DAWs are capable of working with sample rates as high as 384kHz at bit depths of 32 bits, when the CD was first invented there were major limitations for both storage and transmission of digital information. This led to the 44.1kHz/16-bit standard for the CD.

The standard of 44.kHz was chosen because it could be easily handled by the only unit capable of recording digital audio at the time, which was a modified professional-grade video-tape recorder (see "Obsolete Formats" later in the chapter for more details).

This frequency was also high enough to provide a 22kHz frequency bandwidth (remember the Nyquist Sample Frequency Theorem back in Chapter 2?), which would cover the range of human hearing.

The 16-bit standard was chosen because it provides a potential dynamic range of 96dB (6dB per bit x 16), which was considered substantial enough for a satisfying musical experience. There was also a limitation of A/D and D/A convertor technology at the time, and 16 bits was state of the art.

The original CDs could store 650MB of information, which is the equivalent of just over 74 minutes of music. This length was determined by the then-president of Sony, who wanted to be sure that a CD could hold the entirety of Beethoven's Ninth Symphony.

Later versions of the CD increased the storage space to 700MB, which extended the playing time to nearly 80 minutes.

HOW CDS WORK

A CD is a plastic disc 1.2mm thick and 5 inches in diameter that consists of several layers. First, to protect the microscopically small pits (more than 8 trillion of them) against dirt and damage, the CD has a plastic protective layer on which the label is printed.

Next there's an aluminum coating that contains the ridges that represent the digital data and reflects laser light. Finally, the disc has a transparent carrier through which the actual reading of the disc takes place. This plastic is part of the optical system (see Figure 12.1).

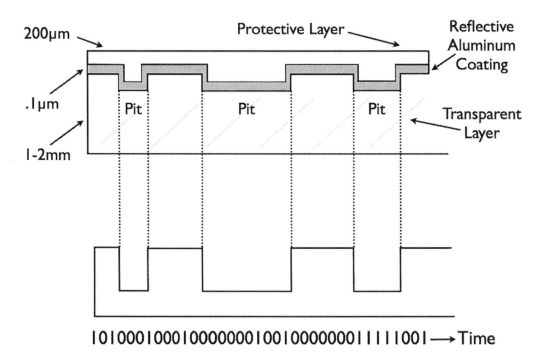

Figure 12.1: The CD has several layers. Notice how the ridges contain binary information

Mechanically, the CD is less vulnerable than a vinyl record, but that doesn't mean that it can be treated carelessly. Since the protective layer on the label side is very thin (only one ten-thousandth of an inch), rough treatment or granular dust can cause small scratches or hairline cracks, enabling air to penetrate the evaporated aluminum coating.

If this occurs, the coating will begin to oxidize, which can cause drop-outs, glitches, or distortion in the audio.

The reflective side of the CD is the side that's read by the laser in the CD player. People tend to set the CD down with the reflective side up, but it's actually the label side that's the most vulnerable, since it's not as well protected as the reflective side.

That's why it's best to store the CD in the jewel case, where it can be safely held by their inside edge. It's been estimated that CDs that have treated well have a life expectancy of between 50 and 100 years.

TIP: CDs are easily scratched and should only be cleaned with a soft cloth wiping from the center to the outside edge, not in a circular motion. Even a small smear running along the grooves can cause information loss when the CD is read, potentially leading to audio dropouts.

The area of the disc that contains data is divided into three areas (see Figure 12.2):

- **The lead-in** contains the table of contents and allows the laser pickup head to follow the pits and synchronize to the audio before the start of the program area. The length of the lead-in is determined by the number of tracks stored in the table of contents.

- **The program** area that contains the audio data and is divided into a maximum of 99 tracks.

- **The lead-out** contains digital silence or zero data and, marking the end of the CD's program area.

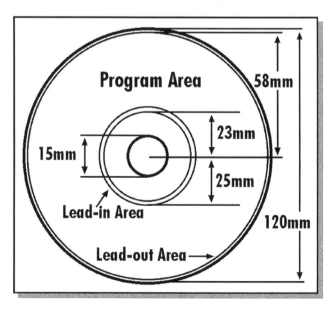

Figure 12.2: The CD layout

Scanning the Disc

Like vinyl records, the information on optical discs is recorded on a spiral track in the form of minute indentations called *pits* (see Figure 12.3). These pits are scanned from the reflective side of the disc (this makes them appear as ridges to the laser) by a microscopically thin red laser beam during playback. The scanning begins at the inside of the back of the disc and proceeds outward.

During playback, the number of revolutions of the disc decreases from 500 to 200rpm (revolutions per minute) in order to maintain a constant scanning speed. The disc data is converted into electrical pulses (the bit stream) by reflections of the laser beam from a photoelectric cell.

When the beam strikes a land, the beam is reflected onto a photoelectric cell. When it strikes a ridge, the photocell will receive only a weak reflection. A D/A converter converts these series of pulses to binary coding and then back to an analog waveform (see Figure 12.4).

Figure 12.3: An electron-microscope look at CD pits and land

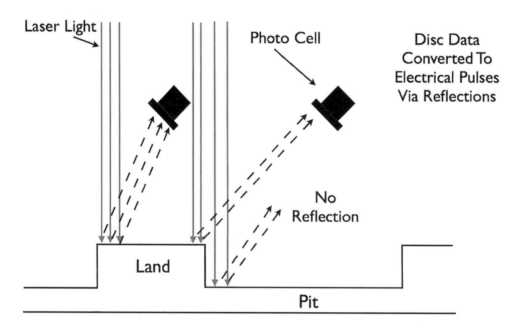

Figure 12.4: The disc data is converted into electrical pulses (the bit stream) by reflections of the laser beam off a photoelectric cell

The ends of the ridges seen by the laser are 1's, while all lands and pits are 0's. Turning the reflection on and off represents a 1, while a steady state is a series of 0's.

Thanks to this optical scanning system, there's no friction between the laser beam and the disc, so the discs don't wear regardless of how often they're played. Discs must be treated carefully, however, since scratches, grease stains, and dust could diffract the light and cause some data to be skipped or distorted.

This problem is solved by a fairly robust error-correction system in the player that automatically inserts any lost or damaged information. Without this error-correction system CD players would not have existed, as even the slightest vibration would cause audio dropouts, glitches, and distortion.

Many CD players use three-beam scanning for correct tracking. The three beams come from one laser, as a prism projects three spots of light on the track. It shines the middle one exactly on the track, and the two other "control" beams generate a signal to correct the laser beam immediately, should it deflect from the middle track.

MASTERING FOR CD

Mastering for CD requires a number of extra steps compared to what's required for music destined for online distribution. That's because there are a number of technical and creative processes that apply only to CD mastering.

Editing PQ Subcodes

If you've ever tried to export an album's worth of songs from your normal DAW timeline, you know that what you get is a large file that plays like a single song instead of individual songs like we're used to on a CD. This happens because you need a special editing workstation for making CD masters that supports what's known as *PQ subcode editing* to separate the songs out.

PQ PQ subcodes manage the track locations and running times on a CD, enabling the CD player to detect how many tracks are present, where they are, their duration, and when to switch from one track to the next.

Editing software applications such as Hoffa CD Burnmaster, Magix Sequoia Pro, Steinberg WaveLab, Zynaptiq Triumph, and DSP-Quattro, among others, have the ability to place these codes as needed.

When the CD was first developed, it had eight subcodes (labeled P to W), and there were a lot of intended uses for them that never came to pass. Today, the P and Q subcodes are mostly used, although the others can contain other information like CD-Text, which we'll cover shortly.

Most PQ editors also allow a PQ logsheet to be printed out, which is then sent with the master to the replicator as a check to ensure that all the correct data and information has been provided (see Figure 12.5).

> **CD Subcodes**
>
> P Channel indicates the start and end of each track and was intended for simple audio players that did not have full Q-channel decoding.
>
> Q Channel contains the timecodes (minutes, seconds, and frames), the table of contents or TOC (in the lead-in), the track type, and the catalog number.
>
> Channels R to W were intended for digital graphics known as CD+G, but they also contain CD-Text data identifying the album, song, and artist.

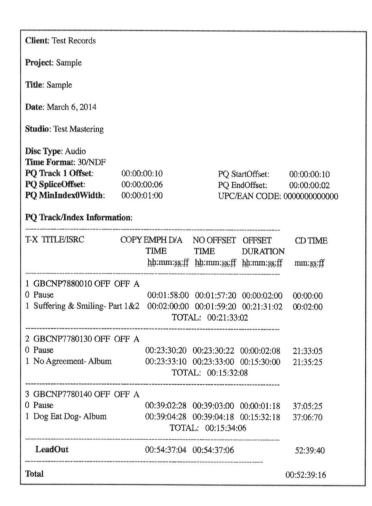

Figure 12.5: A PQ logsheet

Inserting ISRC Codes

Most songs that are commercially released have what's called an ISRC code, which stands for International Standard Recording Code. It's a unique identifier for each track that lists the country of origin, the registrant (releasing entity, usually the record label), the year, and a designation code (a unique identifier created by the label).

This code stays with the audio recording for its entire lifetime. Even if it later appears on a compilation, the same ISRC will accompany it.

If a recording is changed in any way (like an edit or a remix), it requires a new ISRC. Otherwise, it keeps the same number, regardless of the company or format. An ISRC code also may not be reused.

In the U.S., the codes are administered by the Recording Industry Association of America (RIAA), which is a trade organization for music labels and distributors. ISRC codes can help with anti-piracy and royalty collection, though U.S. radio isn't very diligent about using them. There is stronger support for them in Europe.

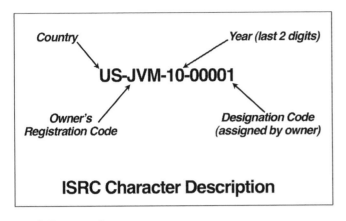

The ISRC is contained in the Q-channel subcode of a CD and is unique to each track. Each ISRC is composed of 12 characters. Figure 12.6 shows what an ISRC looks like and what all the characters mean.

Figure 12.6: ISRC code character description

There can be confusion about when a new ISRC code is required, but most cases are covered in the following list:

- Multiple recordings or takes of the same song produced even in the same recording session and even without any change in orchestration, arrangement, or artist require a new ISRC per recording or take.

- A remix, a different mix version, or an edited mix of a song requires a new ISRC.

- If the playing time changes due to an edit, the song requires a new ISRC.

- Processing of historical recordings requires a new ISRC.

- Re-release as a catalog item requires a new ISRC.

- A recording sold or distributed under license by another label can use the same ISRC.

- A compilation album where the track isn't edited or changed in any way may use the same ISRC.

So how do you get an ISRC code? If you distribute your music through TuneCore or CD Baby, they'll automatically assign one for you. Many CD replicators will also assign ISRCs for you, but they'll charge you a fee for doing so.

However, it's easy enough to register for an ISRC yourself. Go to usisrc.org to register (it will cost a one-time fee of $95), and they'll assign you a three-digit registration number. You can then assign ISRC codes to all your new or previously recorded music that doesn't have an ISRC assigned to it already. Just be sure to keep a good list of the numbers and follow the rules, which are provided on the site.

Inserting UPC Codes

Another code used in the release of most albums is a UPC code. The UPC stands for *Universal Product Code*, which is the number represented by the barcode on the back of the packaging for just about any item you buy in a store (see Figure 12.7).

Figure 12.7: A typical UPC code

While an ISRC refers to a single track, the UPC code is for the entire album, and each unique physical product that is put on a store shelf has this unique code. In addition to the barcode that you find on the back of the CD package, you can actually encode this into the PQ information on a CD.

If you plan to sell your CD in retail stores and have it tracked by Luminate (formerly SoundScan) for inclusion on the Billboard music charts, you need a UPC. Most retailers only stock products with barcodes so they can easily keep track of them in their inventory, and Luminate doesn't know you exist until you have a barcode to identify your CD, cassette or vinyl record.

UPCs are administered by the Uniform Code Council and they charge $30 for a single barcode, but you can also get a single UPC at barcodecreate.com for $8.

Inserting CD-Text

CD-Text information includes the album title, song title, artist, and song duration, which are displayed on a playback device if it is CD-Text enabled (note that some older players are not). This data is stored in the R through W subcodes, as is karaoke info, graphics, and other extended features not standard to the original CD spec.

Most applications that allow you to insert PQ codes will also allow CD-Text info to be inserted, but it's not automatic and must be selected in a menu. Once again, only specialized mastering applications allow you to insert CD-Text information with your CD master.

Song Order

You don't have to think about the song order much if you're planning to release your songs online, but the song sequence becomes important when releasing an online album, CD, or vinyl record.

In the case of a CD, a sequence that grabs the listener right at the beginning and orders the songs in such a way as to keep listener attention continually high is the goal, but it's a creative decision, so really anything goes.

Choosing the song sequence is typically the responsibility of the artist and producer, not the mastering engineer, and should be decided before mastering even begins..

That said, it's something that the mastering engineer needs to know before the job can be completed and the CD master delivered.

Adjusting The Spreads

When mastering for CD or a vinyl, the time between the songs is called the *spread*, and it can be used as a creative tool just like the song sequence.

The spreads determine the pace of the album. If the songs are close together, then the pace feels fast, and if they're further apart, it feels slower. Sometimes a combination of both feels about right.

Many times the spread is timed so that the next song will correspond to the tempo of the previous song. For example, if the tempo of the first song was 123 beats per minute, the mastering engineer might time the downbeat of the next song to match that tempo. The number of beats in between depends upon the flow of the album.

Sometimes, a cross-fade is used between songs, eliminating the spread altogether, but this is typically a decision made during mastering.

Many disc-burning utilities, like Roxio Easy Media Creator and Toast, have only limited spread selections, usually in 0.5-second intervals. That should be enough for most situations, but if you need more precision, you'll need a dedicated PQ editor, as discussed previously.

Using Dither

Dither is a low-level noise signal that's added to the program in order to trim a large digital word length into a smaller one. Since the word length for an audio CD is 16 bits, audio with a longer word length, such as the 24 bits commonly used in DAWs, must be reduced.

Simply cutting off the last 8 bits (called truncation) degrades the audio quality, so a dither signal is used to smooth this process. A truncated, undithered master may have abrupt decay trails or buzzing distortion at the end of fade-outs and generally won't sound as smooth as a dithered one.

Dither Types

Not all dither is created equal. There are many algorithms for applying dither, with each DAW manufacturer offering their own version or using one from a third party.

Dither generally comes in two main types: flat and noise-shaped. Flat dither sounds like white noise and slightly increases the noise level, while noise-shaped dither shifts much of the noise to frequencies beyond human hearing.

Although it seems like using noise-shaped dither would be a no-brainer, many mastering engineers continue to use flat dither because they claim that it tends to "pull together" mixes. Plus, if it's a loud track, you'll be hard-pressed to hear it anyway.

There are many excellent dither algorithms from companies like Waves, iZotope, and POW-r, as well as plugins like PSP's X-Dither, which lets you hear the dither in isolation. Each offers different dither options for different styles of music.

It's worth trying all the types before settling on one, as each can affect songs differently, even within the same genre.

TIP: When mixing, make sure to turn off the dither in your DAW before exporting your master mix file. Dither should only be applied once, at the very end of the signal chain during mastering, to be effective.

Rules for Using Dither

Apply dither once, and only once. Since dither is a noise signal, applying it multiple times will have a cumulative effect. Plus, dither introduced too early in the signal chain can have a very detrimental effect on any subsequent DSP operations that occur afterward.

Insert dither only at the end of the signal chain. The time to dither is just before exporting your final master.

Experiment with different types of dither. Each type sounds different, and one may suit a particular style of music better than others. However, the differences are usually subtle.

DELIVERY FORMATS

A CD master uses a special industry-standard delivery file format called Disc Description Protocol or DDP. This file can be delivered to the replicator on a CD-ROM, DVD-ROM, Exabyte tape, or FTP file transmission, which is the current standard. DDP uses more robust error correction than audio CDs, ensuring that the audio master received by the replicator has minimal data errors.

The DDP Master

DDP is the preferred master medium for several reasons:

- DDP has far fewer errors than any master medium, thanks to computer data error correction.
- It's easier and safer to go past the 74-minute boundary.
- DDPs are safer. Once a DDP image is created, it can't be altered.

Once again, a DDP image file can only be generated from a DAW app that has true mastering capabilities.

FTP Transmission

Most replicators accept master files via FTP (File Transfer Protocol) and prefer this method of delivery. When using FTP, the best thing to send is a DDP image file, since it already contains the necessary error correction to protect against transmission errors.

All replicators have a secure portion of their website dedicated to FTP transfers. After placing your order, they'll send you the host name, user ID, and password.

Obsolete Formats

Although some pressing plants may still accept recordable CDs as masters, this wasn't always the case. For background's sake, here are a couple of formats that have since been made obsolete by DDP.

The Sony PCM-1630

Time for a bit of history. A longtime staple of the mastering scene was the Sony 1630 (see Figure 12.8), which is a digital processor connected to a Sony DMR 4000 or BVU-800 3/4" U-matic video tape machine. In the early days of CDs, it was the only way to deliver both a digital program and the accompanying PQ information to the replicator for pressing.

The digital output of the PCM-1630 (and its predecessor the 1610) was recorded to a 3/4" U-matic videotape cartridge (see Figure 12.9), which was known for its low error count.

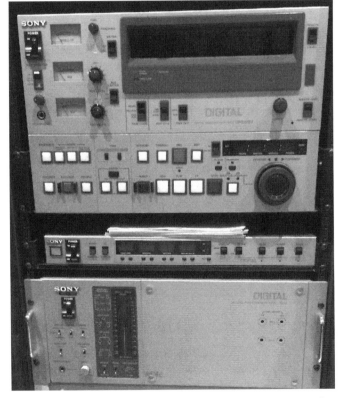

Figure 12.8: A Sony 1630 digital processor with a BVU-800 3/4" machine

If a replicator still accepts the format, the glass mastering (see the next section) from the U-matic tape can only be done at single speed, so the audio data is usually transferred to DDP for higher-speed cutting (which is not necessarily a good thing to do from an audio standpoint).

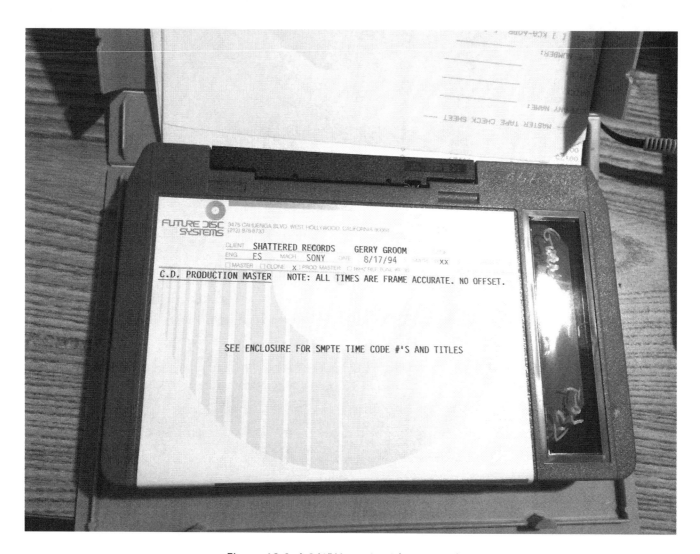

Figure 12.9: A 3/4" U-matic video cartridge

The PMCD

Another relic from the early CD era is the PMCD, which stands for Pre-Mastered CD, a proprietary format co-owned by Sonic Solutions and Sony.

It was originally intended to replace the Sony PCM-1630 as the standard medium for delivery to the replicator. Unlike a normal CD-R, the PQ log is written into the lead-out of the disc (see "How CDs Work" earlier in this chapter). During read-back, this log generates the PQ data for glass mastering, eliminating one step in the replication process.

Though a good idea at the time, PMCD was short-lived because the playback devices went out of production, and much of the equipment required for the process was available only from Sony, limiting the replicators' options. The more modern DDP has proven to be an solid replacement, thanks to its robust nature and superior error correction.

HOW CDS ARE MADE

CD replication is very similar to the process of making vinyl records (which is outlined in Chapter 11, "Mastering for Vinyl") in that it involves multiple steps to make the components that actually stamp out the discs. Here's an overview of how that works.

1. Replication is composed of a number of stages that are required to create a master from which CD stampers are produced. All of the processes are carried out in a Class 100 clean room that's 10 times cleaner than a hospital operating room and has a very low amount of dust particles, chemical vapors, and airborne microbes. To keep the room in this state, the mastering technicians must wear special clothing, such as facemasks and footwear, to minimize any stray particles.

    The replication master starts with 8-inch diameter, 6mm thick glass blanks recycled from previous replications. Preparation begins by stripping the old photo-resist from the surface, followed by washing with de-ionized water and careful drying. The surface of the clean glass master is then coated with a photo-resist layer a scant 150 microns thick with the uniformity of the layer measured with an infrared laser.

    The photo-resist coated glass master is then baked at 176 degrees for 30 minutes, which hardens the photo-resist layer and makes it ready for exposing by laser light.

    Laser-beam recording is where the photo-resist layer is exposed to a blue gas laser fed directly from the source audio of a DDP master tape or file. The photo-resist is exposed where pits are to be pressed in the final disc.

    The photo-resist surface is then chemically developed to remove the photo-resist exposed by the laser and therefore create pits in the surface. These pits then extend right through the photo-resist to the glass underneath to achieve the right pit geometry. The glass itself is unaffected by this process (see Figure 12.10).

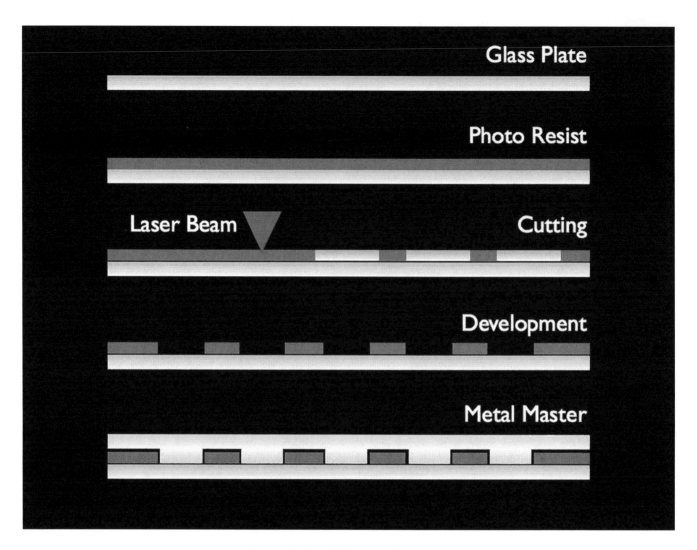

Figure 12.10: Making the CD master

2. The surface of this glass master, which is called the *metal master* or *father*, is then coated with either a silver or a nickel metal layer. The glass master is then played on a disc master player to check for any errors. Audio masters are actually listened to at this stage.

3. The next stage is to make the reverse image stamper or *mother* (a positive image of the final disc pit and land orientation).

4. Stampers (a negative image) are then made from the mother and secured into the molding machines that actually stamp the CD discs.

5. After a CD has been molded from clear polycarbonate, a thin layer of reflective metal is bonded onto the pit and land surface, and a clear protective coating is applied.

6. The disc label is printed on the non-read surface of the disc, and the CD is inserted into a package such as a jewel case, with a tray, booklet, and backliner.

A single unit called a *Monoliner* is used to replicate CDs after the stamper has been created. The Monoliner consists of a complete replication line made up of a molding machine, a metalizer, a lacquer unit, a printer (normally three-color), and inspection.

Good and bad discs are automatically sorted onto different spindles. Finished discs are removed on spindles for packing. It's also possible for the Monoliner to not include a printer, so a new job can continue without being stopped while the printer is being set up.

A Duoliner is a replication machine made up of two molding machines, a metalizer, a lacquer unit, and inspection. Each molding machine can run different titles, with the discs being separated after inspection and placed on different spindles.

> **Duplication Versus Replication: What's The Difference?**
>
> Some people use the terms duplication and replication interchangeably but there is a difference.
>
> Duplication means to make a copy of something, so in the case of a CD it means burning a copy of a disc. A computer extracts the digital data from the master disc and burns it onto a recordable CD, DVD, or Blu-ray disc, making the new disc a copy of the master. The recordable disc was manufactured first, then the information was burned onto it later.
>
> Replication means that the new disc is created from scratch, with the information from the digital master embedded during the manufacturing process. A replicated disc starts as a lump of polycarbonate before being stamped with the data. It's then manufactured with the information already embedded..
>
> There is a slight quality difference between duplicated and replicated discs, although it's minimal these days. A duplicated disc may have errors as a result of the burning process that could degrade the sound. A replicated disc theoretically has none of these errors.
>
> Some 'golden-eared' mastering engineers claim they can always hear the difference between the two, though the average listener likely won't notice..

TAKEAWAYS

- A CD's digital audio resolution is 44.1kHz sample rate with a word length of 16 bits.

- The average CD can hold up to 74 minutes of music, although versions with up to 80 minutes are possible.

- The protective layer on the label side of a CD is very thin, so rough handling or dust particles can cause scratches or hairline cracks. This allows air to penetrate the evaporated aluminum coating, leading to drop-outs, glitches, or audio distortion.

- Optical discs store data on a spiral track as tiny indentations called pits, representing digital 1s and 0s. A microscopically thin red laser scans these pits during playback.

- Creating a CD master requires a specialized editing workstation capable of inserting PQ subcode information.

- PQ subcodes control track locations and allow the CD player to detect how many tracks are present, their durations, and when to switch between tracks.

- Most commercially released songs have an ISRC code (International Standard Recording Code), a unique identifier for each track that includes the country of origin, the registrant (usually the label), the year, and a designation code assigned by the record label.

- UPC stands for Universal Product Code, the number represented by the barcode on the back of packaging for nearly any item sold in stores, including CDs.

- Spreads, the time between songs, can be used as a creative tool just like the sequence of the tracks.

- Dither is a low-level noise signal added to reduce a 24-bit digital word to a 16-bit one for CD use.

- An undithered master may have abrupt decay trails or buzzing distortion at the end of a fade-out, and generally won't sound as smooth as a dithered master.

- Dither should only be applied once, at the very end of the signal chain during mastering, to be effective.

- Disc Description Protocol (DDP) files are the industry-standard format for delivering audio files for CD replication.

13

IMMERSIVE MASTERING

Immersive audio has taken the music industry by storm, with distribution of immersive music now available online from streaming services like Apple Music, Deezer and Tidal, in video games, and on Blue-Ray discs. With it comes a need for mastering, but that also means a whole new set of rules for the engineer, as both immersive mixing and mastering are much more involved than mastering for stereo.

Unfortunately there are multiple immersive formats available that make both production and playback somewhat confusing. While the Dolby Atmos immersive format currently has the widest user base by far, there are other formats available as well, such as Sony 360 Reality Audio, Auro 3D, DTS-X, and L-Acoustics L-ISA.

In this chapter we'll primarily look at Dolby Atmos, but we'll also touch on Sony 360 Reality Audio so you can see the differences between them. But first, immersive audio has a rich history that goes back to the 1940s, and it's worth knowing so you can understand how the field evolved into what it is today.

THE IMMERSIVE AUDIO BACKSTORY

You may think it's new, but immersive audio (or "Surround Sound" as it was formerly known) has actually been with us in one form or another way longer than you might think.

Theatrical releases began using the three-channel "curtain of sound" developed by Bell Labs back in the early 1930s when it was discovered that a dedicated center channel provided the significant benefit of anchoring the center by eliminating "phantom" images (in stereo, the center image shifts as you move around the room). This also provided better frequency response matching across the soundfield as an added byproduct.

The addition of a rear effects channel to the front three channels dates as far back as 1941 with the "Fantasound" four-channel system utilized by Disney for the film Fantasia, and then again in the 1950s with Fox's Cinemascope, but it didn't come into widespread use until the 1960s, when Dolby Stereo became the de facto surround standard. This popular film format uses four channels (left, center, right, and a mono surround, sometimes called LCRS) and is encoded on two tracks.

For many years, Dolby Stereo was the standard delivery format for all major shows and films produced for theatrical release and broadcast television as it had the added advantage of playing back properly in stereo or mono if no decoder is present. This makes it compatible with a wide variety of both new and old theater sound systems.

With the advent of digital delivery formats capable of supplying more channels in the 1980s, the number of rear surround channels was increased to two, and a low-frequency effects channel was added to make up the six-channel 5.1, which soon became the modern standard for most films, surround music, and digital television.

Today we've graduated to far more advanced formats highlighted by the totally revolutionary multi-speaker Dolby Atmos system, which debuted in 2012.

And then there was the four-channel Quad from the 1970s, the music industry's attempt at multichannel music that ultimately failed due to two incompatible competing standards, both of which suffered from an extremely small sweet spot. Needless to say, that's a lot of different formats.

The LFE Channel

Before we look at the different immersive speaker formats in detail we must understand the concept of the LFE, or "Low Frequency Effects" channel.

All immersive playback formats designated as "x.1" (like 5.1 or 7.1) offer this dedicated subwoofer channel (the ".1" is the subwoofer), which is sometimes referred to in film production circles as the 'Boom' channel, as it's designed to enhance the low frequencies of a film. This provides that seat-shaking effect during earthquakes, plane crashes, explosions, or any other dramatic scene that calls for lots of low frequencies.

The LFE channel, which has a frequency response of about 30Hz to 120Hz, is unique in that it has an additional 10dB of headroom. This extra headroom is necessary to handle the power required to reproduce low-frequency content without distortion.

Bass Management

Subwoofers in these systems also include what's known as a bass manager (sometimes called bass redirection), a circuit that takes all the frequencies below about 80Hz (this frequency may vary) from the main channels, along with the signal from the LFE channel, and mixes them into the subwoofer (see Figure 13.1).

The reasoning is that, since the playback system already has a subwoofer, it's beneficial to use it for more than just occasional low-frequency effects. This allows the effective response of the entire playback system to extend down to about 30Hz, provided the subwoofer is large enough.

This is not necessarily needed in a Dolby Atmos system however, since all channels can be full range and may not require the bass extension from the subwoofer.

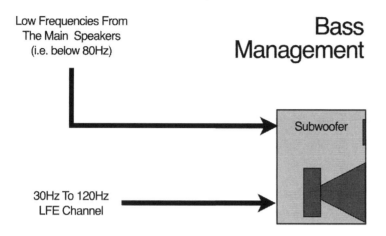

Figure 13.1: The LFE channel.
© 2024 Bobby Owsinski, All Rights Reserved.

Since the majority of consumer surround systems (especially the inexpensive ones) include a bass-management circuit, if you don't monitor with one, you may not hear things the way people at home do. And, since the bass manager provides a low-frequency extension below that of the vast majority of studio monitors, the people at home may actually be hearing things (such as unwanted rumble) that you can't hear while mixing.

> **TIP:** *In various configurations, the first number designates the number of main channels. If a low-frequency effects channel is included, the format is labeled 'x.1' (e.g., 5.1); if not, it's labeled 'x.0' (e.g., 4.0).*

1st Generation Immersive Audio Formats

Immersive audio can be divided into three distinct eras, with the first two generations being more "surround" in nature rather than fully immersive. This is because, for the most part, they were based on audio coming directly from the available speaker channels (you'll see the differences soon).

The first generation consisted of the "x.0" formats, or the ones not using a subwoofer. All the formats in this generation were also delivered in analog. This generation includes the three channel (3.0) Dolby Surround, the four channel (4.0) LCRS, the four channel (4.0) Quadraphonic, and the five channel (5.0) Dolby Pro Logic.

Various Speaker Arrangements

Dolby Surround was an encoding format that used stereo front speakers and a mono speaker in the rear for surround (see Figure 13.2). It was first introduced in 1982 as a way to encode surround information onto CDs and VHS videotapes.

LCRS stands for Left, Center, Right, Surround and is the basic cinema setup of three speakers behind the screen in front of the audience, each fed by separate channels, along with a single mono surround channel that feeds multiple speakers throughout the theater.

The first true consumer surround format, called "Quadraphonic" (or just "Quad" for short), was introduced in the early 1970s. It consisted of a stereo pair in front of the listener and another stereo pair behind them. The format never caught on primarily because there were two competing delivery methods, causing consumers to hesitate in purchasing one or the other for fear of choosing the less popular standard, which might not be supported (see Figure 13.3).

Dolby developed its Pro Logic encoding specifically for delivery of multichannel audio to the home. Gradually the number of channels it could encode evolved from three-channel (stereo with a single surround), to four-channel (LCRS with a single surround), and finally to five-channel (LCRS with stereo surrounds). This became a popular format for television audio, especially when it became possible to deliver it digitally.

3.0
Dolby Surround

Created For
Home Use

Figure 13.2: 3 channel Dolby Surround.
© 2024 Bobby Owsinski

Figure 13.3: Quadraphonic surround
© 2024 Bobby Owsinski

2nd Generation Immersive Audio Formats

The second generation consisted of the "x.1" formats which incorporated a subwoofer and were also the first delivered in a digital form. This included formats delivered via DVD, Blu-Ray disc, or electronic transmission for television. Second generation formats also experimented with more channels in an attempt to achieve a higher degree of realism.

However, the main problem wasn't in the production of the content, but on the consumer side, where the speaker systems were rarely positioned or calibrated properly in the home so the effect was disappointing as a result.

Even More Speaker Formats

The six-channel 5.1 surround was the gold standard for many years in both theater sound and home theater. The format consists of three front channels (L,C,R), two surround channels (left surround or Ls, and right surround or Rs), and a low-frequency effects channel (LFE) (see Figure 13.4).

5.1 Surround Created For Home And Cinema Use

Figure 13:4: 5.1 surround.
© 2024 Bobby Owsinski

The 5.1 format dates back to 1976, but became the de facto standard for surround sound upon the arrival of the DVD in the early 1990s. Although many other formats were attempted (as you'll see), they all stem from the basic 5.1 setup. Only now is Dolby Atmos beginning to surpass the format in both usage and awareness.

Although 5.1 was the standard for theater surround sound for a long time, many film mixers found the format too limiting, as it made it difficult to localize effects with only two surround channels. As a result, the 6.1 format was created to offer a center rear surround channel in an effort to improve the localization.

As listener tastes became more sophisticated, the eight-channel 7.1 added surround channels on the sides for greater immersion (see Figure 13.5).

For a brief time, Sony offered its own eight-channel format known as "Sony Dynamic Digital Sound," or SDDS. This was a 7.1 format that was configured somewhat contrary to what you might expect. Instead of additional surround channels, five channels were used across the front. One reason for the format was that it provided the higher sound pressure level needed for larger theaters. However, many also argued that it offered increased panning precision across the front speakers.

Tomlinson Holman originally coined the term "5.1" while he was the chief scientist at Lucasfilm, and his TMH Labs took surround much further than the rest of the industry with his 10.2 format. This is similar to the 7.1 format except for the addition of stereo LFE channels, a center rear channel, and stereo height channels placed in the front over the screen. 10.2 was actually a precursor to the immersive audio setups that we use today.

As an aside, Tom is also responsible for the THX sound standard that you see in movie theaters; it stands for "Tom Holman eXperiment."

To further immerse the listener in sound, the 12-channel 11.1 system was developed. This format adds additional side surround channels as well as stereo height channels above the screen.

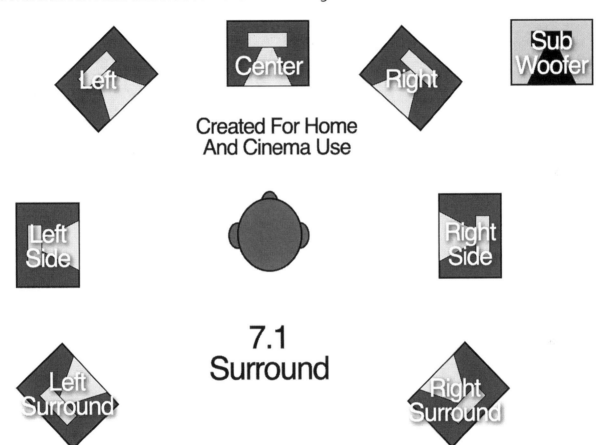

Figure 13.5: 7.1 surround.
© 2024 Bobby Owsinski

All of these formats were based on sound coming out of speaker channels, so panning was always on a single horizontal plane. Even with height channels, there was limited ability to pan in three-dimensional space due to the available production tools, which was one of the first challenges addressed by third-generation immersive formats.

3rd Generation Immersive Audio Formats

Surround sound truly became immersive audio with the introduction of the multichannel Dolby Atmos system in 2012. With a whole new set of digital production tools and true overhead height speaker channels, mixers were finally able to create an audio experience that earlier generations could only dream about. Today the capabilities are so vast that the only true limitation is the mixer's imagination.

Just as in the Quad days, multiple technologies are competing for the same consumers, with the addition of Sony 360 Reality Audio, Auro 3D, DTS-X, and L-Acoustics L-ISA (although L-ISA is mostly used for concert and theater sound). So far, Dolby Atmos has a significant lead in installations, available tools, and song releases, so that's what we're going to focus on here.

> You might be wondering, "Why is immersive audio so different from previous surround formats?" The biggest difference in 3rd generation systems is that panning shifted from channel-based to object-based.
>
> This means that a mixer is no longer panning towards a speaker but instead into a three-dimensional space in the listening area. With more speakers in use (up to 64, although that's primarily used in theaters), these objects can envelop the listener in sound more seamlessly and naturally than ever before.

Unlike second-generation immersive formats, integration into the typical home is now much easier and more attractive. Now a simple soundbar, a subwoofer, and a couple of wireless speakers would not only get the seal of approval from the family worried about the decor, but provide a very enjoyable listening experience as well.

AN INTRODUCTION TO DOLBY ATMOS

A Dolby Atmos playback system takes listener immersion to a whole new creative level. That said, it also has a number of additional advantages beyond the creative ones, such as:

- It's both future-proof and backwards-compatible with previous 2nd generation formats.
- It's easy to monitor between a wide variety of immersive formats.
- It can also be played back in regular stereo on non-Atmos-enabled devices.

Although the system supports up to 128 channels routed to 64 speakers (which is most often utilized in theaters), the typical music production space utilizes a 7.1.4 (as seen in Figure 13.6) or 9.1.4 playback system. That means a 7.1 or 9.1 system with the addition of 4 height channels positioned in a square above the listener. This speaker configuration coupled with the addition of the Dolby Production Suite panner allows the mixer to place an object anywhere within a 3D space in the room.

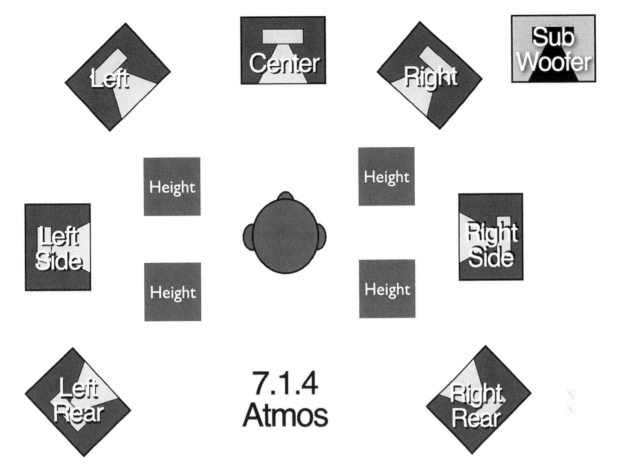

Figure 13.6: A typical 7.1.4 immersive speaker setup
© 2024 Bobby Owsinski

Beds And Objects

While mixing in first- and second-generation formats involved bussing or panning sounds between speaker channels, third-generation immersive formats use a different concept called 'object-based mixing.' This means you still have the traditional panning to channels as before, but you can also pan other mix elements throughout a full 3D space. These are what's known as beds and objects.

Beds

Atmos, like other immersive formats, still allows you to pan in a horizontal plane towards any speaker that you'd like, plus it also adds two overhead channels. This is known as a Bed. The Bed audio can range from stereo up to 7.1.2 (with the .2 representing the Left and Right Overhead channels).

While you can have multiple Beds if you desire, most mixers just use one that's oriented more towards the front of the room. Some call this the "front wall" of the mix.

> **TIP: In music production, Beds are particularly effective for low-frequency information, as the only way to assign a mix element to the LFE channel is through a Bed. Mix elements that work well in a Bed include individual drum tracks, percussion instruments, bass, lead vocals, and reverb.**

However, some mixing approaches forgo the Bed entirely and rely solely on Objects, which we'll explore next.

Objects

Objects have the unique ability to pan anywhere within 3D space, rather than simply toward the speakers. Objects can be panned in three coordinates - an X plane (left to right), a Y plane (front to back) and a Z plane (up to down). Even better, the size of that space can also be easily controlled with a simple Size parameter control (see Figure 13.7).

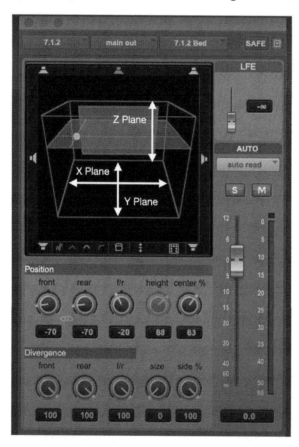 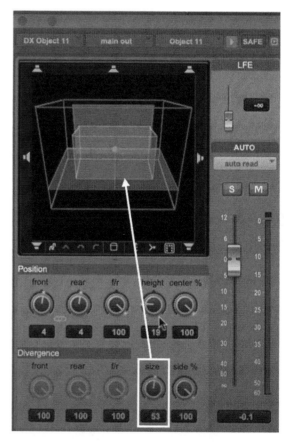

Figure 13.7: The panning and size parameters
© 2024 Bobby Owsinski

Objects are typically used for the precise and dynamic placement of mix elements such as synths, guitars, percussion, and effects.

A huge piece of Object-based mixing is that all of the size and positional data is captured as metadata, which will be used in various ways later, as you'll soon see.

Rendering

Atmos can be confusing because it's actually a two-part system consisting of the Atmos panner and another essential processor called the Renderer (Dolby refers to it as the Renderer Mastering Unit, or RMU). Instead of routing to a 2-channel or multichannel output bus in your DAW, you route up to 128 audio channels to the Renderer.

The Renderer is actually the place where all the various output formats can be created simultaneously, and where the audio is monitored. This takes place during both content creation and consumer playback (see Figure 13.8).

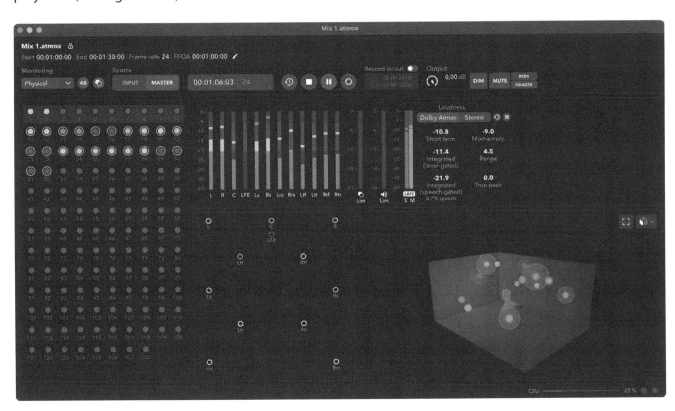

Figure 13.8: The Dolby Atmos Renderer
Courtesy Dolby Laboratories

During mastering the Renderer can be hosted on the same computer as the DAW software (it's now built into many DAWs), but for large sessions like in film post production, it's more efficient to host it on a separate computer.

Regardless of how the Renderer is hosted, one of its main functions is to allow you to listen to the mix in any number of formats without needing to patch or reconfigure anything. That means you can hear the mix in 7.1.4, or 5.1, 3.0, or stereo by just selecting the appropriate output.

Pro Tools, Logic Pro, Nuendo, Cubase, Studio One, Pyramix, and DaVinci Resolve have the renderer built into the software, though the feature set is not as complete as with the full Dolby Renderer, which must be purchased separately.

Unlike with Generation 1 and 2 playback systems, Generation 3 consumer playback systems are intelligent, and combined with the Renderer, provide a superior immersive experience even with few speakers available.

The Renderer built into in a smart consumer device will sense the speaker configuration available, then derive a format from the Atmos file that best reproduces the mix in that environment. Devices like the Apple TV4k, Roku, Amazon Firestick, and Amazon Echo Studio support Atmos, and services like Apple TV+, Netflix, Amazon Prime, and Max currently stream the format.

Binaural Playback

Since the vast majority of consumers will be listening to an Atmos version of a song on headphones, the Binaural Renderer processes all of the Bed and Object audio to create a compelling immersive mix over headphones using head-related transfer function (HRTF) filters.

This replicates the experience of listening to immersive audio on speakers as closely as possible while using headphones. A mix created while monitoring via loudspeakers will also translate to the binaural renderer.

Unlike the Renderer output meant for speakers, the Binaural Render has three mode settings that let you determine the perceived distance of each object and bed. These settings are OFF, NEAR, MID, and FAR, and there's a big difference between them.

TIP: It's best to monitor the Dolby renderer's binaural mix with all distance settings set to MID first before finishing your binaural mix.

Immerse Virtual Studio
A more advanced way to monitor an immersive project is by using the Embody Immerse Virtual Studio, either the Alan Meyerson or Lurssen Mastering editions. The Alan Meyerson edition provides a virtualization of famed film mixer Alan Meyerson's Studio M, while the Lurssen Mastering edition presents a virtualization of the immersive mastering studio used by mastering engineers Gavin Lurssen and Ruben Cohen.

The Apple Music Conundrum

Atmos song versions are now available on multiple streaming services like Amazon Music, Tidal and Apple Music. Amazon and Tidal currently use the Dolby Atmos codec as you hear it from the Renderer, so your mastered project will sound true to what you heard while mastering. Apple Music is a different story however.

Although Apple Music also utilizes the Dolby Atmos codec, the platform uses its own Spatial Audio Visualizer that's different from Dolby's. This provides an immersive experience, but may also present a substantially different representation of the song.

Many artists and engineers are surprised when they hear what their mix sounds like on the platform, so it pays to be aware of the differences before a mix is sent to the service.

The reason why mixes sound different on the platform is that Apple's Spatial Audio decoding algorithm ignores the binaural distance settings of the Dolby Atmos encoder.

Also be aware that Apple Music may play a Spatial Audio mix up to 10dB louder than other platforms, making it more likely to trigger the playback limiter and change the sound as a result. You can prevent this from happening by setting your iOS device to Soundcheck ON for the best experience.

A good way to monitor Apple Spatial Audio in your DAW is by using the Embody Immerse Virtual Studio Apple Music Signature Edition.

The Loudness Limit

Atmos mixes have a fixed target loudness of -18LUFS. While this might seem considerably lower than modern stereo mixes, there's a good reason for this setting.

Since Atmos is great at folding down a 7.1.4 mix to fewer channels (even down to stereo), the summing of the 12 or 14 immersive channels to 2 will naturally increase the overall level. Staying at -18LUFS or below provides enough headroom so you'll never have to worry about an overload occurring after a fold down.

Because Dolby Atmos allows for the full frequency range in all speakers, it's important to be aware of where bass-heavy Objects are panned. Excessive low-end in an Atmos mix may take up too much headroom and cause a mix to be rejected as a result.

Atmos Files

There are several types of files associated with an Atmos mix. Before we get to them, be aware that the resolution for a Dolby Atmos mix can be up to 96kHz/24 bit and that's what's encoded into the Atmos session file. The sample rate of the tracks in your DAW can be different (like 96kHz for instance), but eventually the sample rate will be transcoded to 48kHz for streaming distribution.

The file that your DAW uses during production is a .atmos master file. In other words it will be named "xxxxx.atmos." This appears on your computer as a folder with usually three or four files inside containing the audio and the metadata.

When it's time to deliver your master, export the project as an *Audio Definition Model Broadcast Wave Format*, or ADM BWF file. ADM BWF files contain a multichannel audio file with the surround bed, individual mono, and split stereo object tracks, along with all the metadata required for reproducing the mix on Dolby Atmos-compatible systems or devices.

A smaller MP4 file can also be generated for use on Apple devices to check the mix using Apple's Spatial Audio codec, or consumer playback equipment, like a Blu-ray player or soundbar.

TIP: While it's true that a stereo mix can be derived from your Atmos file, you still need to create a separate stereo master for Apple Music as well. The reason is that your stereo master runs in parallel with your Spatial Audio mix during playback. This allows the listener to switch between the two formats on-the-fly, but it also requires that they be tightly synchronized.

The Renderer also allows you to bounce your Atmos mix to other discrete surround audio file formats as well if required.

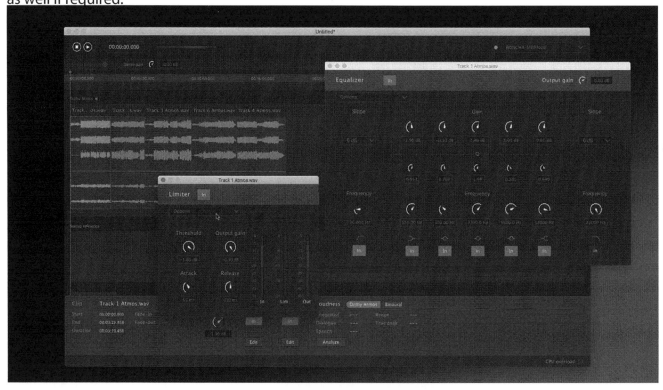

Figure 13.9: The Dolby Atmos Album Assembler
Courtesy Dolby Laboratories

The Album Assembler works with Dolby Atmos ADM BWF files and treats each one as a single block very much like different takes of a song on the same DAW timeline. In other words, it doesn't provide access to each individual bed and object channel. This simplifies the workflow and brings the experience closer to what mastering engineers typically work with in stereo music.

Stereo Synchronization

For each ADM BWF file, the Album Assembler displays a 5.1 rendered waveform block representing each song. The Album Assembler also includes a stereo reference track so that you can align your Dolby Atmos song starts and durations with mastered stereo content.

TIP: All Atmos ADM BWF files must match the length of their corresponding stereo clips.

Monitoring is done through the Dolby Atmos Renderer application, and the Album Assembler manages all metadata, including any trim, downmix, and binaural settings in the source ADM BWF file.

TIP: The stereo reference track can also be monitored, but since the mastered stereo content is typically louder than Dolby Atmos content, you'll probably want to lower its level.

The Album Assembler lets you adjust the gain, EQ, and limiting with the processing applied equally to all channels in the ADM file. It's important to note that the limiter is not a brick-wall limiter like what's typically used in a mastering stereo program. This means that it's best to watch the meters in the main application window and Renderer to be sure that you're not clipping or creating too much binaural limiting. You can apply the same EQ and limiter treatment to multiple songs by just copying and pasting the settings between clips.

Once you've finished mastering, select the Dolby Atmos clip and export it as a mastered ADM BWF file. All trim, downmix, and binaural metadata is now encoded in the ADM export.

Loudness Analysis

One critical but often overlooked element of the Atmos Render is the Loudness Analysis tool. This will not only provide two critical loudness parameters that are needed for your submitted master to be accepted by a streaming service, but also allows for final level adjustments as required.

Follow these steps (Courtesy of Alex Solano):

- Access the Loudness Analyzer and click the Analyze Loudness button. It will then show you the Integrated Loudness level, Integrated Dialog level, Speech Percentage, Loudness Range, and True Peak level.

- Be sure that the Integrated Loudness level is below -18LUFS. For example, a reading of -17.6LUFS is closer to 0dB, so it would be too hot and require the mix level to be lowered.

- Be sure that the True Peak level is below -1dBTP. If it's -.8dBTP, it's too close to 0dBTP, and the file will most likely be rejected.

- Next go to the Trim and Downmix Controls window. Set the 5.1 and 5.x Downmix settings to Direct Render.

- Go to the Trim Controls, set them to Manual, then make sure that the Surrounds and Height are set to zero. This ensures that if your Atmos mix is downmixed to stereo, the levels of all objects will stay true to your original mix.

SONY 360 REALITY AUDIO

Major label acceptance has driven the adoption of immersive audio, but there is some division as to which format is preferred. For instance, Warner Music and Universal Music are both firmly behind Dolby Atmos, while Sony Music prefers the format originated by its parent company, Sony 360 Reality Audio (360RA).

Atmos - 360RA Differences

Atmos and 360RA are quite different and not compatible with each other, so it helps to understand the differences between them.

Speaker Layout

One of the biggest differences is the speaker layout. 360RA uses 13 speakers as standard as compared to the 12 speaker Atmos setup. However, the positioning of these speakers is quite different.

360RA uses a 5.0.5+3 setup, which means 5 main speakers, with 3 across the front and two in the rear, similar to a 5.1 setup of the past. There is no subwoofer (hence the 0 in between the 5's).

It also includes 5 overhead speakers positioned above each of the main speakers, with 3 across the front and 2 in the rear.

Probably the most unusual aspect of the setup is the 3 bottom speakers (the +3 in the designation), positioned across the front set up below each of the main horizontal plane speakers (see Figure 13.10).

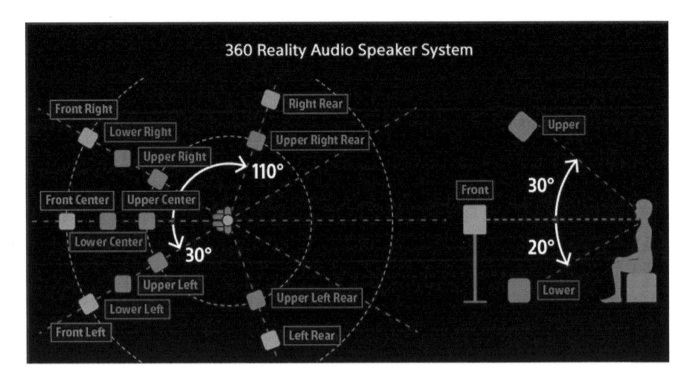

Figure 13.10: Sony 360 Reality Audio speaker layout
Courtesy Sony Electronics

Channels

While Dolby Atmos supports up to 128 channels, 360RA is limited to 24. As a result, Atmos is capable of a more immersive experience.

While its true that almost anything beyond 9.1.4 is used primarily for theatrical releases, having additional channels also allows for greater panning possibilities.

Other Differences

Both Atmos and 360RA are object-based, so there are no differences there, but Atmos playback is available on a wider range of devices. Also, because Atmos is free to manufacturers while Sony charges a fee for 360RA use, the adoption rate is much lower.

As mentioned earlier in the chapter, immersive audio mastering is more complex than stereo due to the different workflows, required equipment, delivery specifications, and varying formats. We are still in the early days of immersive audio, so expect these processes to become much easier as time goes on.

TAKEAWAYS

- At the time of writing, multiple immersive audio formats exist—Dolby Atmos, Sony 360 Reality Audio, Auro 3D, DTS:X, and L-Acoustics L-ISA.

- Dolby Atmos is the most widely adopted format by creators, record labels, music distributors, and consumers.

- The speaker layout for Dolby Atmos and Auro 3D is 7.1.4, while Sony 360RA is 5.0.5+3 (with 3 bottom speakers).

- All immersive formats use height channels. In a 7.1.4 setup, 4 height speakers are arranged above the listener.

- The LFE (Low Frequency Effects) channel represents the '.1' in 5.1, 7.1.4, etc. It typically rolls off at 120Hz and uses a bass manager to provide low-end extension through the subwoofer alongside the main channels.

- Older formats like 5.1 and 7.1 are channel-based, meaning you can only pan toward specific speaker channels.

- All immersive audio formats are object-based, meaning that you can pan a mix element anywhere in a 360 degree space. Some formats like Atmos are both channel and object-based.

- Atmos can have as many as 128 channels, but anything beyond 9.1.4 is primarily used for cinema playback.

- Atmos delivery files are in Audio Definition Model Broadcast Wave Format (ADM BWF).

- All Atmos ADM files must be delivered at a maximum level of -18 LUFS with a True Peak level below -1 dBTP.

- Immersive audio is always played back as a binaural track when using headphones, so the binaural settings are critical for accurate reproduction.

- Your immersive track may sound different on Apple Music compared to Amazon Music or Tidal, as Apple uses its own codec, which differs from Dolby's. Be sure to check your master on each platform.

- The only dedicated mastering tool for Atmos is the Dolby's Atmos Album Assembler.

The Mix Fix Playbook
15 Quick Solutions To Your Most Common Mix Problems

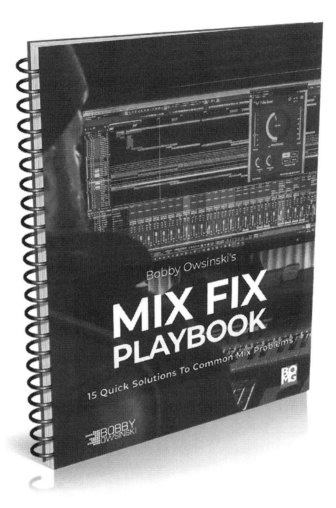

Find out more at <u>go.bobbyowsinski.com/mixfix</u>

(Sorry, due to high shipping costs this offer is only available for customers in the United States, but you can still get it on Amazon)

PART II

THE INTERVIEWS

14

MAOR APPELBAUM

Maor Appelbaum has worked with the likes of Faith No More, Yes, Meatloaf, Eric Gales, Walter Trout, Dream Theater, Sepultura, Halford, and many more. He's written presets and collaborated with various plugin companies such as Waves, Brainworx, Plugin Alliance, Softube, Arturia, Leapwing Audio, Pulsar Audio, IK Multimedia, Newfangled Audio and others.

He's also lectured at various trade shows and recording schools, and is the co-creator of The Oven line of analog hardware, which is also available as a plugin by Brainworx and other companies.

How long did it take before you felt confident mastering?
There was definitely a point where I felt comfortable and confident, but then came projects that were more challenging, which made me reevaluate. It's like climbing a mountain. You think you're almost at the top, and then you hit a big bump that makes you realize you're not quite there yet.

How confident you feel really depends on the type of projects you're working on. Some projects require a lot of changes and intrusiveness, while others just need subtle moves. Sometimes when you're used to making big adjustments, you suddenly get a project that only needs minimal changes, and you have to rethink your approach.

Early on, your confidence is often tied to your setup—your gear, your room, your monitoring situation. I changed a lot of gear and worked in different rooms over time. I improved my monitoring, and now, yes, I feel confident in my work today.

But that doesn't mean I wasn't capable before; it just means that every challenge I faced taught me something new and I took all that moving forward. New situations bring new approaches and that's part of the growth that adds to your ability and confidence.

Each project is a new learning experience with its own challenges. It's not just about how you hear the music—it's about how the client hears it too. You might think you've made the best possible decisions, but the client might steer you in a different direction.

At the end of the day it's the music they had created and you helped them achieve their sonic signature. That being said, It still has to work for them in the long run

Is there a type of music that you find more challenging to work on than others?
That's a really good question because I think every genre of music comes with its own set of ideas and challenges. For example, when you're working on fast-paced music like metal, the low end can easily muddy up the sound because the BPM is so fast and the kick drum hits are very quick with lots of low end build up.

In those cases, you need to keep the sound tight as much as possible unless they want it to be more loose and sluggish. On the other hand, in some other genres, the low end can bloom and be round and smooth. They're so different from each other.

Having experience with various types of music is crucial, but it's a longer path to learning. If you focus on just one genre for a long time, you can get really good at it, and that could be your calling card. Once you start branching out into other genres, every new style presents a new and fresh learning curve.

Every genre—and even sub-genres—has its own aesthetics. Some are softer, some are brighter, and each has its own specific characteristics.

Just because a track is nearly perfect doesn't mean it's easy to work on. In fact, those projects can be the hardest because everything is placed so well that even a slight frequency adjustment can shift the inner balance between instruments.

At the same time, dense mixes that aren't perfect can also be challenging. You want to improve them, but there are limitations. When you start fixing things, you might solve one problem but create another.

For example, if a mix is too muddy, with too much low end, or even too much high end, you try to clean that up. But in doing so, the mix might lose its energy and power. You fix the original issue but introduce a new one.

The challenge with mastering a stereo track is that any changes you make are both improving and degrading the track at the same time. You're fixing problems by altering the sound, but by altering it, you're also losing some of its original fidelity.

Sometimes that's a positive thing because it sounds better overall, but other times, you can end up doing more harm than good. It's tricky, it's about picking your battles and a balancing game

Is there a common problem that you hear in mixes that come into you?
I think nowadays, the range of issues is broader than it was in the past. Years ago, there was often too much low end because people were working in rooms that were far from being treated and tuned or that didn't have enough clarity in the low-end, or they were using speakers that couldn't handle it. They'd boost the low end in the mix, and then you'd end up with mixes that were overloaded with low frequencies that needed cleaning up.

The issue wasn't just the low end, though. When they boosted the low end, it often masked the high end, so they'd compensate by boosting the high end as well.

Once you cleaned up the low end, the high end would suddenly pop out too much. These kinds of problems are generally caused by monitoring issues, especially in small or untreated rooms.

This is likely to be a issue now thanks to the new room tuning softwares and other apps they can check the sound on and reference. More people have treated rooms, better speakers, and improved setups. Producers have a much better starting point with current DAWs, plugins, hardware, and conversion.

Another common issue is when artists work on a project for a long time and get ear fatigue. They can end up over-processing things without realizing it.

For example, you might get a mix that's too saturated, and not in a good way, just because they didn't notice things becoming too harsh, abrasive, or bright. Sometimes it's about the levels—like the cymbals might be too dark while the vocals are sibilant, or the vocals are dull and the cymbals are too bright.

These imbalances are often due to monitoring, but they're also affected by how long someone has been working on a project without taking breaks or referencing other songs.

Rushing a project can have similar effects. You might end up printing a track with an issue because you didn't catch it during the process, especially when using offline bouncing.

I often hear issues in tracks where automation hasn't worked properly, or something jumps out that shouldn't. You'll sometimes find a verse that's really quiet and then a chorus that's way too loud, or the other way around, or maybe automation cuts off unexpectedly.

This happens especially when there are hundreds of tracks and tons of automation running.

What do you ask people to send you? Is there a certain level you want, or do you ask them to take the limiter off or anything like that?
The first thing I ask producers or engineers when they send me a track is whether they've made a loud version and a non-loud version. If they say, "Yes, we mixed into a limiter or a mastering chain," I ask them to send me two files—one with the limiter in the master chain and one without.

The same applies if they've made a version for the client using a limiter, multiband compressor, or other processing. I ask for both versions at the same sample rate and bit depth, so the quality is consistent across both files.

I also ask them to include any fade-ins or fade-outs, so I know exactly how it's supposed to sound.

Years ago, I would ask for a certain level because I noticed that people were often pushing things too hard and it didn't sound good.

Back then, DAWs didn't handle hot mixes as well, and some plugins couldn't deal with those levels effectively. Nowadays, DAWs have better headroom, and plugins are much more capable of handling hotter signals, so now I just ask for one version with the processing and one without.

Sometimes, I also request screenshots of the master bus. Certain projects use specific processing, and if I hear something strange in the track, looking at the master bus can help me identify if a plugin is causing the issue.

When you get a project in to master, do you know what it's going to sound like in the end?

My approach can be broken down into two phases. First, I listen to the mix before I do any work on it. At this stage, I'm just trying to identify any potential issues that I can flag for the mixer or leave it as is.

Sometimes during that initial listen, I don't hear any glaring problems, and that's when I remain neutral. I don't try to envision anything yet—I'm just listening objectively to see if anything jumps out or if something is off in the speakers.

Once I start working on the track, new problems can emerge, especially when I start raising the volume, adjusting the EQ, or doing other processing.

For instance, I might find that there are too many sub frequencies triggering the limiter, or maybe some high-end frequencies are causing issues with de-essers. At this stage, it's more like a science project—I experiment with different gear and techniques to see what works best with the track.

Even if I have an initial idea of where I want to take it, that's not the full picture. I think about how to make the track groove better, move nicer, how to get the energy flowing, how to make it sound calm and smooth if needed, or give it a cozy vibe. Sometimes I aim for a hi-fi sound or a more immersive feeling—not necessarily Atmos or surround sound, but more of an "in-the-room" presence.

This part of the process is very experimental for me. And, surprisingly, I sometimes think something is going to work in one direction, but it turns out the track needs to go in another direction entirely. It's all about reactiveness—pushing the sound in a certain direction and seeing how it reacts.

I wouldn't say I have a 100% vision for every track right from the start though. I can only predict where I want to take it, but the real proof is in the process. I experiment until I get to the point where it feels right.

Sometimes it's really good, but not quite there yet. Other times, it's too much. It might take longer, but by going through this process, I can test different approaches and eventually find that perfect balance which really hits it nicely

What kind of masters are you typically delivering?

For every project I take on, the first thing I ask is, "What are the output formats?" I still do CDs because there are artists who continue to release them, so that's often part of the project.

For CDs, I provide the DDP format to the client to send to the pressing plant, along with WAV files and MP3s. Some projects also require high-resolution files, M4As, or other variations of MP3s. Vinyl prepped files are a big part of the process too, as some artists want files specifically optimized for vinyl.

Interestingly, cassettes are also making a bit of a comeback, so I prepare cassette sides. There are also projects for DVD and Blu-ray, especially for bands releasing live shows, and I do a lot of audio work for those live recordings as well.

Do you care about LUFS levels, and do you do any streaming masters?

I'm still not sure that LUFS (Loudness Units relative to Full Scale) are as reliable as they should be because I use different LUFS meters, and they're often not even close to each other.

Sometimes the readings are so far off that I don't know which one to trust. Occasionally, they get closer, but I don't fully understand why. I'm saying this not because I lack knowledge—I just haven't had anyone come in and show me definitively how this is supposed to work.

I've mastered tracks that show very high LUFS levels, and they sound the same as those with lower LUFS levels. I've even reduced compression, and the LUFS stayed the same. So, I honestly can't tell you which LUFS target to aim for because every meter gives me a different reading.

I use both hardware and software LUFS meters, and I've even heard some LUFS meters can change the actual sound of the program when you run it through them, which is crazy. I'm not the only one who's noticed this—it's one of those issues that can happen with plugins.

In the end, everything I work on ends up on streaming services anyway. Even projects that started as CDs or vinyl releases eventually get streamed. I don't view the LUFS numbers for streaming as a hard target. Most tracks I've seen aren't even close to the requested numbers, and I don't think they need to be.

If I master something and it sounds good at a certain volume—regardless of the LUFS target—that's where it should stay. Lowering the volume affects the sound, even if you're just turning down the final level. Any adjustment to the processing changes the sound. If streaming platforms change it, it's out of my control.

That's why I'm skeptical about rigid target levels and high-res standards. Streaming services apply their own algorithms, and each one has different targets. So, in my view, the best thing to do is give streaming platforms as few reasons as possible to make changes.

Let's talk about gear for a second. What speakers are you using?

For the past twelve years or so I've been using PMC IB1S passive speakers, driven by a Bryston 4B SST^2 amp with Academy filters between the amp and the speakers.

My monitoring chain goes through a Maselec MTC-1X console, with the monitoring converter being a Lynx Hilo. My main converters are JCF Latte DA/AD and Lavry Quintessence DAC / Savitr ADC for all playback and capture.

I have other converters such as the JCF AD-8 ADC, Crane Song HEDD DA/AD, Rupert Neve Designs MBC ADC, and Benchmark DAC1 DAC. The Lynx Hilo feeds the Maselec MTC-1X monitor section going into the Bryston amp, and then the PMC IB1S speakers. It also works as an interface between the DAW and my system routing.

In my current room, I'm using a Trinnov system for room correction, but for nearly a decade before this, I didn't use any electronic or electroacoustic processing.

You're not fully "in the box," right? You're also using analog hardware?
Yes, 99% of my work is done externally using hardware gear.

What workstation are you using?
I divide the mastering process into two sections: pitch and capture. The pitch is sending audio out of the system through the gear, and capture is bringing it back in. For this, I use a DAW—typically Pro Tools. I send the audio out of the computer through the system, and then I choose the pitch-capture converter.

This could be a JCF Latte, JCF AD-8, Lavry Savitr, Lavry Quintessence, or one of the other converters I have. I do all the processing with hardware, and then I bring the audio back into Pro Tools for capture.

When I'm working on singles, this is a straightforward process. For albums or EPs where I have more tracks to work with, I print each track individually and then sequence them afterward.

Sometimes I'll sequence directly in Pro Tools if needed, but in other cases, I'll use a dedicated sequencing program. It could be an older program like WaveBurner, or something like WaveLab—essentially a two-track editor designed for sequencing.

Once I've sequenced the tracks, I either create a DDP directly from the program, or I export the sequence to another program to create the DDP for the CD. I like to split the work between two or sometimes three programs if there's something specific I need to achieve.

You've designed some of your own gear. How did that come about?
Over the years, I've used a lot of gear from different manufacturers, and all of it played a role in the projects that I worked on. There were some pieces that I liked, but I wished they worked a bit differently, so I started modifying them.

When I began modifying and improving certain units, I started getting ideas for new equipment, but I needed to figure out how to make them. Eventually, I got in touch with builders who helped me create custom gear based on my ideas.

Later on, I decided to develop gear that was completely my own, not just modifications of existing equipment. I worked with Roger Foote from Foote Control Systems, and together we created three units: The Workshop, The Bench, and The Torch.

These are unique, one-of-a-kind units that I have. We thought about producing them for others, but the parts were hard to source, so we kept them exclusive to my setup.

Afterwards, I collaborated with Chris at Hendy Amps, whom I'd been in touch with for years. Together, we started working on an idea called "The Oven," (see Figure 15.1) which was originally designed as a custom unit just for me. Eventually we had something that sounded so good, we thought, "Why not turn this into a product?"

Figure 15.1: The Oven

Later, we developed additional units like The Stove, The Cooker, and The Grill, each with a different tonal signature and purpose. These units complement each other, and you can combine them to achieve a variety of sounds.

What do they do?
They're essentially tone shapers, with some incorporating EQ while others focus more on certain tonal ranges like high-end or low-end.

They also add a level of saturation, but not in an overly aggressive way. They have a wide range where you can find the sweet spot—it's not like there's nothing happening below a certain point and too much above it. You can really play with them to enhance depth, size, punch, and impact.

I use them frequently, and because they offer different sounds and options, I sometimes cascade them, running one into another, or even two or three in sequence. Most of my rig today is made up of custom gear, though I also use a few pieces that are either modified or remain as they are. I mix these with the custom gear to create the best results.

The Oven line of products—The Oven, The Stove, The Grill, and The Cooker—have even found their way into the live rigs of major tours like Linkin Park, The Killers, The Red Hot Chili Peppers, Smashing Pumpkins, Party Next Door, CCR, Beck, Megan Thee Stallion, Joe Satriani, Sammy Hagar, and Bruno Mars. It's amazing to see big shows running their mixes through this gear.

The Oven has its own plugin version, developed in collaboration with Brainworx and Plugin Alliance. It's a successful plugin now, used all over the world by big-name artists, and it's applied on everything from movies and TV shows to songs.

That said, I still prefer working with hardware because I enjoy how the interaction between different pieces of gear creates unique results. I don't mind using hardware and writing down all the recall sheets. It's part of the "no pain, no gain" process, and I love it.

The thing with hardware is that it allows you to make a conscious choice. If you have both hardware and plugins, you're making a deliberate decision to use one over the other, not because you lack something, but because one works better for the song.

I have plenty of hardware and plugins at my disposal, and I choose whichever works best for the project at hand. It's about having the choice, not about being limited by what's available.

Of course, I'll add plugins if necessary, but I rarely work fully in the box. I've done it, and it can be done really well, but for me, it's a very small percentage of the work.

Do you have a philosophy about the gear you choose?
I treat everything as an EQ and a transient shaper. Whether it's a compressor, an eq, a limiter, a converter, a console or any other audio analog or digital processor.

To me, they all shape the sound in both the frequency balance and the transient response. A compressor doesn't have to compress to do its job—it can act as a tone shaper and just gel the sound based on the sonics of the circuit itself.

I often use compressors for their sound, not just for the compression. The same goes for converters. I'll pick different DAs and ADs based on how they color the sound, because, to me, they are all some kind of EQs without being labeled as such.

Are you considering any immersive work at all?

Currently, I focus mostly on stereo work, which comes naturally to me since it's how I've grown up in the industry. I don't see myself getting into Atmos for the time being. That could change—never say never—but I think the platform still has a long way to go in terms of production, mixing, and mastering.

Right now, stereo and mono are more than enough for me. I say "mono" because there are still some projects in mono that come my way, and I enjoy working on them.

I feel there's always more to explore within stereo.

What's the best piece of advice that you would give to a young mastering engineer?

I believe you can learn a lot from your clients, often more than from watching videos or other educational resources. Their feedback—whether they like something or not, and how you respond to what they don't like—teaches you a lot in the long run.

The more work you do, the more you learn and gain experience, which I think is more valuable than just learning specific techniques. Techniques might get you from point A to point B, but truly understanding what a client wants can take you from A to Z.

15

ERIC BOULANGER - THE BAKERY

Eric Boulanger is the founder of The Bakery mastering studio as well as a professional studio violinist.

A protege of legendary mastering engineer Doug Sax, Eric has mastered GRAMMY-winning or nominated projects for Green Day, Hozier, Selena Gomez, Colbie Caillat, OneRepublic, Imagine Dragons, Neil Young, The Plain White T's, Chris Botti, and many more.

A classically trained violinist since the age of three with over two decades of professional experience, Eric has trained at such renowned institutions as The Juilliard School, Manhattan School of Music and Tanglewood. His performances and recordings range from orchestral and chamber music, Broadway musicals, contemporary pop music, and film/television scores.

Besides all that, Eric is truly unique in that he has an EE degree so he really does know where all the electrons are going, and he's one of the few newer generation mastering engineers who cuts vinyl as well.

Bobby Owsinski: When you begin a mastering job, can you hear the final product before you start?
Eric Boulanger: Yeah. The first thing I do is listen to what has come through the door. I'm a bit unconventional because I don't really rely on meters or anything like that. I know everything is fine-tuned to such an extent that I just trust my ears.

Before I even touch anything, I'm already thinking about the sound I want to achieve. That process happens instantly for me. I'll sit and listen, thinking about how I want to approach the project. I'll come up with different plans—A, B, C, D—and then start executing and tweaking things as I go.

What's interesting is that music can surprise you. Sometimes, I'll have a clear concept of exactly what I want, but when I start executing the plan, it doesn't work as expected, and I'll need to go back to the drawing board.

But to directly answer your question, I always know exactly what sound I'm aiming for—it's instantaneous for me. It's just the method to get there that can sometimes surprise me.

Is there a common problem with the mixes that come to you?

Well, there are many examples, but let me get philosophical here and quote my first professional mentor, the person who introduced me to [mastering legend] Doug Sax and the world of mastering— Al Schmitt. He wasn't a mastering engineer, but we all know him.

I was an intern at Capitol, and I was doing odd jobs and trying to wedge myself into sessions. Then Al started to notice me and thought, "This kid's all right" and began bringing me into all his sessions. Over time, I started as a fly on the wall and gradually got to do more and more. One day, we had a mix session in Studio C, and I ended up running Pro Tools for him, just the two of us in the room.

Since we were alone, and Al was near the end of the mix, he said, "All right, kid, feel free to ask any question you want." Within a minute, I was firing off questions like a machine gun, and Al quickly grew tired of it. He turned around and said, "Eric, you think too much. It's simple—just get it right the first time. Now stop asking me questions." Then he went back to work.

But that moment was a lightbulb for me. The philosophy of "get it right the first time" stuck with me, and it's so applicable to mastering. There are a million ways to approach things, and sometimes people fall into the mindset of "we'll fix it in post." But in mastering, the final step in the process, you're often the one left dealing with everyone else's mistakes.

The philosophy of getting it right the first time should carry through every step of the process—from the first note played to delivering the album. If the sound coming out of the speakers isn't what you want it to be, you need to address it right then.

A particular example of this is stem mastering. You often have artists or less experienced people ask, "Do you do stem mastering?" without fully understanding what they're asking for. It's not that I never do it—there are times when it's necessary for specific reasons – but the idea of taking a mix, stemming it out, and delivering it to me feels like undoing the work you've already set out to do.

If a mixer sends me stems expecting me to turn it into something that matches their vision, my response is, "Why aren't you doing that now?" It doesn't make sense. When people ask whether I prefer working with stems or stereo mixes, of course, I prefer stereo. The stereo mix represents what everyone has been hearing and what they've been working toward. Ideally, I shouldn't have to do much at all to it.

The best case for a mixer is when they don't have to touch anything either. It's that simple "get it right the first time" philosophy, and Al was absolutely right about that.

Is there a certain type of music that you find either easier or more difficult to master?

I work on everything, and I hate being pigeonholed. For mastering, having a wide range of projects is so important. It keeps things fresh and exciting. If I were doing the same thing every day, it wouldn't be as fulfilling. But this is just personal to me—being a violinist, the easiest thing for me is working on projects with orchestras, movie soundtracks, or even pop tracks where there's a solo violinist.

The moment I hear the violin, it's immediate—I can't quite explain it. The EQ settings or whatever adjustments I need to make just come to me right away, even down to the specific numbers. It's like I have a built-in reference point because I know exactly what a violin should sound like.

So for me, working with violins is exceptionally easy, but everything else tends to follow the same general process.

How long did it take you when you first started mastering before you felt that you were good at it?

That's a great question, but one I can't directly answer, and I'll tell you why. It was probably around two years after I started working for Doug Sax at The Mastering Lab. I was his assistant, running computers, adjusting knobs, doing recalls—all that kind of stuff. I was really into the technical side of things, too. In fact, I built the vinyl room we were using.

I didn't initially plan on becoming a mastering engineer. My goal was to follow in Al Schmitt's footsteps and become a recording engineer or producer.

As time went on, I started learning more and more, and I really fell in love with the process. I enjoyed mastering because it's about quantity as well as quality. I work on an album a day, while mixers often work on an album for a month. That pace really suited me. Plus, it allowed me to continue playing the violin professionally, which I loved.

After about two years working with Doug, I had a realization: I would be an idiot if I didn't ask him to train me as a mastering engineer. So I asked him, and he said, "No problem, but you need to find a paying client." He didn't care about the rate, but I had to get a real client and let them know I was starting out, with Doug backing me up in case anything went wrong.

Amazingly, within 12 hours, I had a client. Doug was stunned.

That's when he gave me the reins. He said, "Go ahead and do it." One of the most important lessons in this industry is developing the confidence to press the button and send the final product to the client when you think it sounds good. Of course, you need the technical skill and musical knowledge, but having the confidence to say, "This is finished" is the hardest part.

That night, I sat in the hot seat with Doug next to me. I was incredibly nervous, especially since it was my first time starting a project from scratch. It took me forever to get through the first track, but eventually, I turned to Doug and said, "I think I'm ready." He listened, said it sounded fantastic, and told me to finish the album. Then he went home.

That was the last time I ever saw him in the studio with me. At first, I thought it was crazy, but then I realized he was showing me that I already had the confidence and the skill. He didn't need to hold my hand. It was a bit like how my dad taught me to swim—he just threw me in the pool and said, "There, you're swimming now!" It was tough love, but it worked.

Okay, so that being said, you have a reference point of what you're hearing and understood how everything was working. Given the fact that there's some very powerful mastering plugins and AI online mastering, which doesn't take into account any of that, how do you feel about it?
I'm not nervous or bothered by this at all. Let me explain with a story. The reason I opened my studio on the Sony Pictures lot is because, as a musician, I wanted to be in a creative, collaborative environment.

Financially, it would have made much more sense to buy a house, convert a garage, and keep all the money, rather than paying Sony Pictures rent. But the thought of working alone in a garage would have been unbearable for me—I would have felt so lonely and disconnected.

Sony Pictures is, of course, a movie lot, but I chose that location because I wanted to be surrounded by creativity. Even though I'm working on records, it's the creative energy of being in that environment that matters.

When you're making a record, you're still making music, even as an engineer. There's artistry involved, a feedback loop just like when you're playing in a recording session. Disconnecting yourself from that creative process is how you end up with music that feels soulless, and nobody wants to listen to that.

So, to get back to the AI question, using AI in music is a massive disconnection from the creative process. AI sees things in terms of right and wrong, but in music, there is no absolute right or wrong.

Sometimes you purposely make a record sound bad because it has a certain feel, and that's exactly what people enjoy. AI wouldn't be able to understand that because it tries to process everything as binary—either correct or incorrect.

I'd love to see AI get revision notes and adapt creatively, but I don't see that happening anytime soon. If it ever does, I'd probably just use it myself to speed up my workflow and take more vacation time. But ultimately, AI is just a tool that assists in the process, not something that can replace it entirely.

Do you ever master anything for a certain LUFS level?
No, with the sole exception of broadcast-specific projects where I have to follow a spec sheet. In those cases, yes, I adjust accordingly. But even then, when I'm working with those specifications, I set up my room to reflect the specs, because spec sheets are often surprisingly different from one another.

I adjust my listening environment to match those specs with my room, but after that I go about things as I normally would because I trust my system and environment. It's not always precise, but my ears get me incredibly close to where I need to be.

For something like a Netflix sync, where they have very tight specifications, I've never had to make more than a half dB tweak to meet those requirements.

Tell me about your monitors and your signal path.

First and foremost, I'm really honored and lucky to be at Sony Pictures, as I mentioned earlier. The room I use used to be one of their smallest screening rooms on the lot. The room where I cut vinyl is where the projectors used to be. The main room, where I do all my mastering and listening, used to be a theater.

One of the happy accidents of this setup is that, typically in studios, the ground is flat and they hang diffusers or clouds above you. But in my room, the ceiling is flat, and the floor is angled like a theater or stadium seating. So, essentially, the diffusion is flipped upside down, but it still works the same way. That's one of the cool quirks of the room.

My main monitors are ATC 150s, and I use a hybrid setup. To break it down: I use Pro Tools HD to bring in sources and for playback into the console or for routing.

Most of the incoming sources go through Pro Tools, and at the end of the chain, we still use Sadie for the record unit. That might change soon since Sadie hasn't been updated in a long time, but I use it because it's always been an excellent-sounding DAW, especially for manufacturing, metadata, and other deliverables that Pro Tools or Logic can't handle.

In between, I use a variety of plugins. I do a hybrid mix, and my favorite plugin is the Massenburg EQ. If you're starting out and need one thing, start with that. From Pro Tools, it goes through the console and the DAC [Digital to Analog Converter]. Both my DAC and ADC [Analog to Digital Converter] are custom-designed by Josh Florian at JCF Audio. We worked on them together to streamline the console routing and avoid unnecessary connections. Instead of using a patch bay with extra connection points, we designed it to be more efficient.

From the DAC, it goes to my line amp, which gives me the option of either a solid-state or tube amp—both are based on the original Mastering Lab design. After that, I have two custom EQs that are my own inventions, based on old UA designs.

My main analog EQ is the Manley Massive Passive, and for compression, I use LA-2As with Kenetek optocouplers that were custom-matched for me. They're specific slow ones, and the amps inside my LA-2As are transformer-less, identical to the ones from the Mastering Lab. They're not the off-the-shelf models.

Any of those inserts lead to the ADC, again custom-built by Josh. One of the cool features we added is that I can switch between analog and digital sample rate conversion (SRC) with matched levels. So, I can easily compare the sound of a digital SRC to an analog chain (D-to-A-to-D), or even a hybrid version, just by flipping a switch. This setup lets me make real-time decisions without guessing whether analog sounds better than digital—I can directly compare them.

A great example of this was a Diana Krall record I worked on with Al Schmitt. People were blown away when they found out it was completely digital because it sounded so good. Diana thought it sounded better that way, too. It became a big audiophile topic, especially when it came out on vinyl.

People assumed it had been processed with all this analog gear, given Al's reputation and my training with Doug. But in reality, there was probably more analog circuitry in their phono preamps than in the entire mastering process of that record.

I would have never expected that myself, but I can tell you that it sounded better.

What kind of masters are you delivering?

When it comes to deliverables, the standard has really become 44.1kHz/24-bit for stereo. I'm proud to say 24-bit, especially when you think about the whole "Mastered for iTunes" initiative. Apple branded it in their typical way, trying to make it sound fancy, but the real achievement was moving the industry away from the old CD standard. Because of Apple's influence, we finally shifted away from 16-bit masters as the norm.

For a long time, you'd burn a PMCD at the mastering studio, send it to the label, and then they'd have someone, maybe even an intern, rip it and upload that to iTunes. I remember thinking, "What are you guys doing? We can use a higher bit depth!" Finally, years later, we're seeing the decline of 16-bit masters. They're still requested for CDs, but for everything else, 24-bit has become the standard.

Of course, it could always be better, but a lot of these aggregator or distributor systems that handle label uploads are pretty archaic. They deal with such massive inflows of data that they can't just flip a switch and upgrade to a new format overnight.

So, for now, 44.1kHz/24-bit is the standard. I see a lot of 48kHz for things like music videos or YouTube, and when I'm working on live concerts or film projects, it's often 48kHz too.

Then there are the high-res requests. I would caution people, though, especially when it comes to Apple. They ask for higher sample rates, and people think that will yield a better result, but in reality, Apple just sample-rate converts it down anyway, so what's the point?

For true high-res releases, we do still get those requests, but SACD seems to have disappeared. I haven't seen one since before COVID, actually, now that I think of it.

Do you do a separate master for vinyl?

So there are two common scenarios when it comes to vinyl. The first is the easier one: I didn't master the record, and it comes from another mastering engineer. They've finished the record, but they don't cut vinyl, so they send it to me to cut.

Sometimes, I'll need to go back to them with suggestions, but this depends on the music. While almost every album nowadays gets a vinyl release, not all music is well-suited for vinyl.

If the music itself isn't a great fit for vinyl, or the master is slammed with compression and loudness, I'll explain that it might not turn out the way they're expecting unless we make some changes. But generally, I handle the intricacies of vinyl on my end.

For example, the biggest issue with vinyl is *de-essing*—it's the Achilles' heel of vinyl records. I'll do that work to ensure it sounds good and then cut the record.

The second scenario is if I did the mastering myself. I can recall everything I did—EQ, compression, etc.—and cut the record directly from the original session. This skips an extra generation in the audio process, which is a big advantage. It's one of the perks of having everything under one roof, and it's a good reason to hire me for both mastering and vinyl cutting.

Additionally, with this approach I have full control over the levels and how hard I want to hit the record based on the length of the material. It's far better to adjust things from a first-generation standpoint. If you make a print and then realize you need to tweak something as simple as the levels, you're stuck manipulating a second-generation file, which isn't quite as audiophile-friendly.

How has knowing how to cut vinyl helped in your mastering? Does it change your approach?

For me personally, cutting vinyl really tied together a lot of deep engineering concepts. I have an EE degree, and in my mind, working on vinyl—especially with the lathe—is like a constant battle.

The lathe fights you, and because of that, it links my engineering side with my musical side. It's almost a perfect combination. Think about it: it's the music coming in, the technical aspects under the hood, and then you're literally creating physical vibrations to make it all happen.

For me it changed a lot of how I approach things, though I'm not sure if it would be the same for everyone else. In general, the main difference—and the best part of this resurgence in vinyl—is that you're not just selling a product. People are freaking out over this old format because it's an experience.

The beauty of vinyl is that artists, producers, and clients come in with a vision, even if they have zero technical knowledge. Some don't even own a turntable, but they come in with a clear idea of how they want their songs ordered and how the album should be presented.

Vinyl kind of demands that level of thought. It makes them focus on creating an experience for whoever's going to be listening, and I think that's the biggest shift we're seeing.

What do you think is the difference between a great mastering engineer and someone who's just good or just starting out?

I think the key thing is confidence. When you have a lot of experience, it becomes second nature. That's not to say someone just starting out won't get there, but the difference is that someone new will second-guess themselves constantly and probably take 200 times longer before showing their work to a client.

When you come in and see me work, I do things quickly and almost mindlessly because I've built up that confidence over time. I know what I'm doing, and I trust my instincts. I can just say, "Yep, this is it," and move on—and people love it.

You only get confidence through time and experience. It's a feedback loop. When you send something out and the client loves it, saying, "Oh my God, thank you, this is amazing," it builds that confidence even more.

16
COLIN LEONARD - SING MASTERING

It takes a lot to convert hit-maker mixers like Phil Tan and Dave Pensado into big fans, especially if you're not in one of the media country's media centers, but Atlanta-based Colin Leonard and his proprietary mastering technology has managed to do just that.

With 13 Grammy nominations and a Grammy win, Colin's credits include Beyonce, Justin Bieber, Jay-Z, Bad Bunny, Migos, John Legend and many more, Colin is proving that there's a new way to look at mastering that's equally effective as the traditional techniques. Along with his custom mastering service at SING Mastering, Colin is also the creator of Aria automated online mastering, the latest trend in convenient and inexpensive mastering.

How did you learn mastering?
The way I learned the most was by pulling up mixes and learning it by ear. I would spend every day doing that for a long while. That's the most important thing because even when you learn other people's techniques you still have to find out what is going to work for you. I don't think that I use many of the techniques that I learned from other people that much.

How long did it take you until you felt that you were good at it?
Every day [laughs]. I feel that it's a lot like playing an instrument. You have hills and plateaus. I always felt the same way playing guitar. You reach a place where you're feeling really confident, then you reach another point where you're working on improving.

Mastering is the same thing. There are always those days where you feel like you're doing amazing work and then there are other days where it just doesn't seem like it's happening, so it's a constant work in progress.

The sonic fashion of music constantly changes as well. Masters from 1999 don't sound like the masters from today, and working like that wouldn't fly with artists or labels. It's a constant learning process.

Is there a certain type of music that you find easier or more difficult to work on?
No, I don't think so. I'm lucky in that I get a lot of different genres, but I don't necessarily find certain ones harder than others. The club aspect of some pop or rap music can make it a little more difficult because the DJs have a different loudness perspective, so that can a bit challenging because they really want everything really loud.

Making it loud yet still having the bass stay loud and clean can be a challenge.

I'm told by some of your clients that you manage to get things a lot louder than anyone else. How do you do that?
I have some proprietary processes that I use. I don't really use plugins. Almost everything that I use is analog. I'm always trying to come up with creative ways to get things louder while trying to keep as much apparent dynamics as possible in the masters.

How did you proprietary processes come about?
Out of frustration, really. I would be mastering some club record that needed to be on 11 and I would get it on the verge of distorting yet the client would say they needed it even louder. I was pulling my hair out.

I try to look at the loudness as something creative, like "How can we get around this without distorting?" All the good mastering engineers face the same hurdles, I think.

Isn't it a lot harder to get the same hot level in the analog domain that you can get with digital?
Yes, and no. I think that for me with digital limiters you basically have a glass of water that's already full once it's hitting zero, and then you're boosting the level up and then it's just overflowing.

With analog, depending on how things are calibrated and how much headroom your equipment has, you can operate it in such a way where the analog signal is much louder and clearer.

Getting it back to the digital domain is where everything falls apart [laughs]. That's where you have to work on creative ways to get the same level without causing problems and for me that's the key to making it happen.

I prefer analog to digital because I find it a little more pleasing to my ears. There are some guys doing all digital that sound pretty good, but I prefer analog.

What's your signal path like?
it's kind of a classic setup. I have a playback computer based around the Cube-tec platform that just plays back the files. I don't really do any processing in it. Then I convert to analog, and then I have a Dangerous Music Mastering mastering console that tweaked a little bit, but the stock one is fantastic.

From there it's mostly EQ and not a lot of compression. I only have a couple of analog compressors that I don't use that much. I have 6 analog EQs that are all good at different things. I have some newer high voltage EQs and then some old Neumann cutting EQs that sound really good in the midrange. I have some Manley stuff with transformers and an SPL PQ that I really like. I then go back into a different Cubtec to get back into the digital domain.

I find it interesting that you don't use compression that much.
I don't get using analog compression for mastering. Sometimes it can be nice for softer acoustic pieces or maybe rock stuff, but I think it causes more damage than it helps to songs with transient drums and extended bass notes.

I'm more about transient energy. I concentrate on not screwing up the really fast high-frequency transients since a lot of the energy from the song is there. A lot of times I'll have a compressor or two in line but it won't be doing anything. It's just there for the tube sound. It's more like using a compressor as an EQ.

Isn't the stuff that you get in that's pretty crushed already?
Yeah, totally. Why would we need to crush it any more? At that point you're just losing more dynamic range and that's not what we need.

I do get mixes that are all over the map level-wise though. Sometimes engineers send me stuff that's really open, while others send me stuff that's really compressed. I have to learn each different mix engineer's style, then after a while I know what I'm going to get from them, but it's always part of a learning curve.

Even within pop music, there might be one great mixer that sends me really compressed stuff, and another great mixer that sends me really dynamic stuff.

I think a lot of it has to do with approvals on mixes. Most mixers don't want to send something in for approval on big projects and not have it be at least close to the level of other commercial releases.

Some of the guys that I know that mix quiet and dynamic get a lot more complaints about level. The clients think the energy's not there so they try another mixer. If the other guy makes it louder then they'll lose the client. At the end of the day you have to do what the client wants.

What's odd about competitive levels is if you go to AES and listen to the speaker demos, they'll always use really dynamic material and people will always comment on how great it sounds.
Yeah, I tend to do that to show off my playback system if someone wants to hear it. I'll play something that's less compressed and quieter and I'll just turn it up louder.

There's also something that's going on in playback systems that makes it advantageous to have a louder brighter final product and that's the full range (singer driver) speaker, which is what most people listen to these days.

The speakers on your phone are a single full range speaker, the same as headphones and earbuds, and the same as most computer speakers. Since you don't have a dedicated tweeter, it's just a darker sounding speaker, which has caused productions to get brighter and louder to get as much volume out of those speakers as possible.

What do you supply for vinyl master?
I used to have a lathe and would cut the lacquers, but I don't right now. It's a fun thing to do but it's also really time consuming. You'll go through days where you'll have a lot of problems with production. Maybe the lacquers or the plates got damaged and then you'll have to spend another day cutting new lacquers.

I'll just create 24 or 32 bit files that are a little quieter for the vinyl premaster, and then just lay out each side as a long wave file. If there are 4 songs on one side, it will be all 4 songs connected as one big wave file. I'll also send the CD PQ information along so the cutter knows where the breaks are between the songs.

How do you view spreads these days? It used to be that spreads between songs were really important and you'd time everything out so it felt right between songs. We're in a singles world, so I guess people don't care as much about that now.
I still do. I do a lot of albums and EPs and I spend a lot of time on that. It's kind of an art on its own. Clients like to sit here and do it. For me it's still important, but yeah, if you just have a bunch of singles it doesn't really matter.

I do it in such a way so that the iTunes files use my spacing, so if you buy an album you will get the same spacing that I created. If there's space between the CD markers, I'll use a setting that keeps the that same space on the final file export so that it will carry over into the individual files.

Is there a typical problem that you see with mixes that you get?
I can't name one thing that happens all the time. Most of the guys I work with are really good so I don't have a lot of complaints.

When you think about it, who's to say what a problem is? I try to view it from the aspect of the mixing engineer and the artist making a piece of art. It's their vision, so what can I bring to it to make it better?

If there's something that's really obvious that wasn't supposed to be there, maybe I'll ask in a real cautious way, but I try not to get too technical about it.

What monitors are you using?
I use Duntech Sovereign powered by Cello amplifiers.

What's your take on high end cables?
Oh, man [we both laugh]! I think they make a difference for sure. How much of a difference for the amount of investment I think you need to decide for yourself.

I actually see some of the biggest improvements with cables coming from power cables. It effects certain pieces of gear more than others, and it also depends what the power is like at your facility. I was at another facility where some filtering cables made more of a difference than they do here because the power was worse.

My power here is transformer balanced so it's really clean and I don't get as much of an advantage as I was getting. There are so many variables and I think it's more of a trial and error kind of thing. It's not about getting all solid silver cables or your masters won't sound good though, but I think that good quality cables matter.

How did Aria come about?
I guess some of it was being loaded up with a lot of work and sometimes being a little overwhelmed about how fast people needed things turned around. I'm sure that a lot of busy mastering engineers feel that same frustration sometimes.

The deadlines might not be realistic or even real, but your client sets those deadlines and you have to follow them. A lot of times, if you don't get those projects done on time, you just don't get those projects anymore.

Another part of it was that from mastering so many projects I have a basic setup that I start with, and figured I could automate some of it. It's a unique idea in that we're doing it in the analog domain. This was before I even heard of any of the other online processes.

The tools that I use in the Aria system are all the tools that I use in my normal mastering setup. It comes in digital, we do some scans and snapshots of the audio in the digital domain, then it uses our custom software to play back from one computer.

It then goes through a real high-end D/A convertor and through the whole analog chain, which has automation elements built in, and then it gets recorded again into the digital domain after a real high-end D/A convertor.

From there we can control the exports into the file format that we want. The whole system is hosted in-house, so the servers are here, which makes it really fast. The file goes directly from you to our server here, so it's immediate. As soon as it's mastered it hits your account for download.

How long did it take you to develop the idea?
I had the idea in 2013. I financed the whole thing from my mastering work, so I've been juggling these two things this entire time. It probably could have gone a little faster if I hadn't done that, but I wanted to maintain control of everything so we wouldn't have to do any silly marketing or anything.

Philosophically, aren't mastering engineers opposed to automated mastering?
Yeah, as I've seen online most of them are, but I think Aria is a different product than what I offer at SING Mastering. For one thing, budgets have really dropped. There are some clients that are willing to

pay for high-end professional mastering, but there are also clients that maybe don't have the money to spend on their project, or maybe they need it in like an hour, so I think it fills a void for some clients.

Isn't there some pushback from mixers as well, mostly because they don't believe the final product will be as good as they expect?
I haven't had a lot of that from mixing engineers. Most of my beta testers were really good mixing engineers and they are still using it every single day. If they didn't like it they wouldn't use it.

Part of the idea is that it creates reference level material. Most engineers need to get a level that's competitive with what's commercially available to get approval from the client. I think it fills a void for mixing engineers in that they don't have to focus on the loudness of the track. It's almost like they're handing in a mastered reference.

How loud is it when it comes out the other side?
It depends on the mix. There are 5 settings that go in order of compression, so it depends on what you're going for. A and D and E are very safe settings and they're not pushed hard; B and C have a lot more analog push to them, which can be great on certain types of mixes.

Once you play with it a few times you get the hang of it. Also, once you pick a certain setting you're not married to that setting. You can remaster it at a higher or lower level at no extra charge.

Explain the multiple levels for buying the product.
You can get it on a per song basis or a subscription basis. The subscriptions are on a monthly basis.

For instance, you can buy a subscription for as many as 100 tracks a month if you have a big album to do with a bunch of different versions. That way you'll be able to do it a lot cheaper than what the single rate is.

Where do you think mastering is headed now that we're in this realm of AI and online mastering?
There will always be room for "personal mastering," which is what I do at SING Mastering, but I think it's nice to be able to give another good sounding option that's more convenient and less money.

We live in an Amazon Prime world where people would rather get something done at home at 3AM rather than waiting around to schedule a mastering session, so there's a big convenience factor involved as well.

That said, there's always going to be a market for high level custom mastering.

Mastering Engineer's Handbook - 5th edition | 214

17

PETE LYMAN - INFRASONIC MASTERING

Pete Lyman is a GRAMMY-nominated mastering engineer, and owner of Infrasonic, an audio and vinyl mastering studio with locations in Nashville and Los Angeles. Infrasonic specializes in audio mastering for all formats including vinyl and immersive.

Pete has mastered Grammy-award winning and nominated albums for Chris Stapleton, Tanya Tucker, Jason Isbell, Brandi Carlile, Sturgill Simpson, John Prine, Weezer, Panic! At The Disco, and many more.

How did you get into mastering? How did you learn it?
I was kind of leading a double life—working in IT during the day and playing in my punk rock band at night. My interest in computers led me to start experimenting with early digital audio workstations (DAWs) and mastering.

This eventually led to meeting my mastering mentor, Richard Simpson. One day, I tagged along with a friend who needed to get some vinyl cut, and I was blown away by the process. A few days later, I went back to Richard's shop and begged him for an opportunity, and that's how I got started.

I'm one of the last mastering engineers to enter the field through the traditional route of cutting lacquers. I started around 1999 or 2000 when I opened a recording studio with a small mastering lab next to the studio.

Around 2012, I had shifted from working on avant-garde, punk, and noise records to doing more country music. I had been working with my producer friend, Dave Cobb, since around 2007 or 2008. We worked on Chris Stapleton's first record, Traveler, and after that took off, it made sense for me to relocate to Nashville.

How long did it take you before you felt you got good at mastering? And then the separate version of that is, how long did it take you before you felt good about cutting vinyl?

Let me start by talking about cutting vinyl, because that was the natural progression for me. After work, I would go in and sit with Richard as often as I could to watch him cut records.

Eventually, Richard started letting me run the lathe. I don't really know when I felt like I became good at cutting records. I've cut tens of thousands of sides by now, and I still feel like I'm learning something new every day.

I have to say, vinyl cutting is one of those things that always keeps you on your toes. The more you cut, the more problems you'll encounter. There are so many pieces to the process. After we cut, the lacquers go through electroplating, then pressing, and there are tons of things that can go wrong along the way.

It's not like cutting a DDP (Disc Description Protocol) for CDs, where 99.9% of the time, everything goes smoothly. With vinyl, there are so many variables, and it's often hard to pinpoint where a problem originated.

Are all your projects cutting vinyl then, or are you doing projects that don't require that as well?

These days I'm working on a lot of projects that don't involve cutting vinyl. Honestly, the bulk of my work now is digital audio mastering. At this point I only cut vinyl for the records I master because my mastering schedule is just too packed.

When I'm mastering a project from the ground up, I know exactly how it's going to translate to vinyl. During the EQ process, I take notes and anticipa`te any potential issues for vinyl so I can address those ahead of time.

When you first start a project, do you know what the final product is going to sound like in your head?

For 98% of the material I work on, I have a good sense of where it should go within the first 30 seconds. It's not a big mystery. If you understand the genre and have a good relationship with the producers and engineers involved, that makes all the difference. Most of my clients are repeat clients—producers and engineers I've worked with for a long time—so we've developed a method and understand how to collaborate effectively.

I think the most important part is open communication, especially during the first few projects. Once you've worked with someone long enough, the process becomes almost automatic.

Is there a typical problem that you see in mixes that come to you now?

Something I've noticed a lot lately is an unnecessary amount of harmonic saturation. Don't get me wrong, I love grit, distortion, and overdrive—I have tons of transformer options here, and I'm a fan of that stuff, but I think people tend to overuse it. Once that distortion is baked into a mix, you can't get rid of it, and if the project is going to vinyl, that could cause problems.

I notice this particularly with newer mixers who may not be working in the best environments and aren't sure how their mixes will translate outside of their room. You can hear it right away—the low end is off, or there's an issue with how the mix translates. For example, their low end might be wonky because they can't hear below 40 Hz properly, or there's comb filtering. These issues become much harder to deal with when you pile on extra harmonic saturation or brick-wall the mix unnecessarily.

I don't have any problems with mixers leaving buss processing on their mixes—in fact, I encourage it. I've heard the argument in the past that some mixers shouldn't worry about buss compression and leave it to the mastering engineer, and to me, that's ridiculous. Buss processing is often what gives a mix its own flavor, so removing that before you send it off to mastering makes no sense.

When you mix into a compressor, you're adjusting levels based on how the compressor is reacting. We can't do that at the mastering stage. The same goes for limiting. If you're going to use a limiter, you'd better be confident that your mix is translating well and that you fully understand your room, because it's going to be much harder to fix any issues at the mastering stage.

What level do you like to get mixes at?

When it comes to levels, I always tell people I don't care about the numbers, because they're irrelevant. It's easy for me to turn something down if needed. My main preference is that it's not clipping—unless that's the sound they're going for. Other than that, I don't go out of my way to tell people how to mix their songs.

For example, I work with a lot of engineers who mix on consoles like SSLs. Those mixes usually come in at a good analog level, probably somewhere between -14 and -16 dB. I tend to think those mixes sound great because those engineers pay attention to gain staging, which is something that I think is often overlooked today. As a mastering engineer, gain staging is one of the most important aspects for me, especially when I'm using analog equipment.

On the other hand, some of the best mix engineers in the world send me mixes at -6dB and I'd never ask them to send an unprocessed or "unlimited" mix. Those are creative choices, and they shouldn't be made by the mastering engineer.

I always tell my clients, if a mastering engineer ever asks you to take off your mix bus processing and resend your files, you should fire them immediately—unless you're new to mixing and unsure of what you're doing. I find it unbelievable that someone would ask for that. The bus processing is part of the song. Who am I to say I could do a better job of it than the producer or the band that spent weeks on that track?

Let's talk about your signal path for a little bit. What monitors are you using?

I have a 9.1.4 setup here, and I'm a die-hard PMC user. I work on the original BB5 XBDs—these are the big ones. Each side has dual 15-inch drivers along with mids and tweeters. I also have a BB5 for my center channel.

My Atmos fill speakers are all PMC as well. I believe they're the Ci65 models, which are typically used for permanent in-wall installations. I've got four up top, along with my wide and rear speakers.

How much immersive mastering are you doing?

We've been doing a lot of immersive work lately, and it kind of comes in waves. We got on the immersive train early, and while that had its advantages, there were also some drawbacks.

I wasn't sure what our workload would look like for Atmos mastering, but now, with all the major label work we do, everyone needs Atmos mastering. Granted, some of the work is being released without mastering, but my argument has always been that Atmos absolutely needs to be mastered—probably more than anything else.

We've seen a lot of issues with tracks going up for streaming and getting pulled down almost immediately because they did not translate well in the medium. Immersive audio is still kind of the Wild West. You've got top-level people trying to do it on headphones, but unless you've got a proper 9.1.4 setup like we do, or even a 7.1.4 setup, there's no great way to monitor it. So it's been a bit of a crapshoot, and we're working hard to make sure everyone is hearing the best quality possible.

What workstation do you favor for mastering for stereo?

I'm a die-hard WaveLab guy and have been for as long as I can remember. I was a PC person for a very long time, but now my primary Workstation is a Mac.

Let's talk about your plugin chain. Typically, what would you use?

For a long time, I didn't feel like plugins could come close to the results I achieved in the analog domain, but I no longer feel that way.

What really changed my perspective was realizing that it's not just about the plugin itself—it's about how you use it. We talked about gain staging earlier, and that's what made the difference for me. I realized it's not as simple as pulling in a plugin, like the Maag EQ4, and expecting it to sound exactly like the hardware. It's great, but what really made things click was when I started using the same workflow in the box as I do out of the box.

That means applying the same gain staging principles so I don't overload my plugins. My theory is that it makes sense to use the software the same way you would use the hardware. I'm not going to send a -7LUFS signal into a plugin because if it's emulating hardware, that would be like saturating the electronics.

Once I started developing this in-the-box workflow and applying similar gain staging principles, everything started to fall into place. In WaveLab, for example, if I'm working on an in-the-box project, I don't just pick random plugins. I have a chain—just like I would with my hardware setup.

About 95% of the time, I use that same chain, though I might bypass certain elements. I send roughly the same gain in and out of the chain as I would with my analog console, and that's when things really started to sound right.

Honestly, when you're working with mixes that are already hot and limited, you're doing a disservice by sending them through another round of conversion. Just because people love the idea of using analog gear doesn't mean it's always the right choice.

I love analog, and I use a lot of AAA gear for mastering here. But if something already sounds good and just needs a bit of light EQ, our job is to "do no harm." In those cases, using a clean digital EQ like FabFilter Pro-Q 3 is often better than running it through a piece of gear for the sake of using it.

It's about using the right tool for the job. For a lot of digital audio projects, staying in the digital domain is the best path. Funny enough, it sometimes takes me longer to work in the box than with analog gear, but it's often the better approach.

Every project is different though. Sometimes I'll do an ITB master and then run the same track through my analog chain to compare the two. That comparison helps me decide which approach to take for the entire record.

So what plugins do you favor?

I'm a big iZotope user, and I love the new version of Ozone because it's like having 30 limiters in one. Limiting is often what I start with before I even EQ, so having so much control is amazing.

I'm also a huge FabFilter fan. I use Pro-Q 3 as my main EQ, and I love their de-esser. I'm a big Plugin Alliance and UAD person, too—I use both. One of my favorite EQs is the Knif Soma from Plugin Alliance, which is great because I also have the hardware version.

In terms of other gear or plugins that I reach for, it really depends on what the project needs. For example, if I'm looking for a low end boost, I'll reach for my Bax EQ. If I'm looking to add some breath to the high end of a mix, I'll use the Air Band on the MAAG EQ4. Everything I use is for a specific, targeted purpose.

Are you going for a certain LUFS level?

Honestly, I don't care about LUFS. I'd be lying if I said I didn't look at meters, but for me, I pay attention to VU meters because I think they're a better way to measure perceived loudness. For example, a rock track at -8LUFS might sound really loud and dense, but an acoustic track at the same -8LUFS is going to seem way louder than the rock track. It's all about using your ears.

I'm fortunate to work with artists who don't care about loudness at all. Chris Stapleton, for instance, just wants the record to sound good—he doesn't want it to sound compressed and overly loud. So when we work on his records, it's great because we focus purely on making it sound good.

So for me, LUFS is meaningless when mastering a record. I tell the interns who come through here all the time: if you want the record to sound consistent, focus on the vocal. In 95% of music, the vocal is the most important element. It needs to be consistent across the record. It doesn't matter if one song measures -10 and another -8; what matters is that the vocal sits appropriately in each track. That's what people are really listening to.

How long does it take you, on average, to master a track?

I've been doing this for almost 25 years, so I've gotten pretty quick. If it's a well-engineered track, I can master a three-minute song in about 15 minutes. Of course, there are times when I get stuck or go down a rabbit hole, but I almost always come back to my first instinct.

My approach to mastering, especially with EQ, is based on the fact that most of the time, your first impression is the right one. That's the lightning in a bottle. If you spend too much time obsessing over something, you'll lose that initial spark.

When I'm working on a full record, I listen to the whole thing first without doing too much. I'm just absorbing the vibe, kind of half-active and half-passive, which is how most people listen to music. They're not sitting perfectly between two speakers, intensely focused on every detail.

After that first listen, I'll go through and rough in some EQ adjustments, making notes along the way. I try to stick with my first impressions and apply those to the songs, but I don't print anything yet. I'm just trying to capture that lightning in a bottle. Once I have my initial ideas down, I'll go back in and massage it.

There are so many self-mastering plugins or now online AI mastering services. How do you feel about that?

It doesn't upset me the way it seems to upset some other people. I've heard the arguments about AI, and to me, it's nothing to get worked up about. Sure, it doesn't sound great, but I'm sure some people are getting decent results from it.

For example, if you're a publishing house with 30-second jingles and you just need them at the proper level, then maybe that kind of service is perfect for you. If you're not tech-savvy and don't want to invest in your own software or learn how to do it yourself, it could be a worthwhile option. But those aren't my clients, so I don't worry about it.

Music is about emotion and collaboration. There will always be people that want to collaborate and form relationships. They want someone who understands what they're trying to achieve and who can put a personal touch on it. Those are the people I want to work with anyway.

What kind of masters are you delivering?
A good portion of what we do, especially for label releases, involves delivering in multiple formats.

We're always delivering standard, HD, and Apple Digital Masters. For clients that are doing CD manufacturing, we provide them with a DDP. Vinyl wise, we can either cut lacquers for our projects in house, or if the client plans to press overseas, we provide them with vinyl cutting files. We've also got multiple rooms for our ATMOs/Sony 360 mixing and mastering. Once or twice a year we even need to deliver a cassette master. We've got all the bases covered.

What's the difference between somebody that's really great at mastering and somebody that's maybe just competent?
One of the hardest things about this job is that people tend to do too much.

There's this idea that you have to throw all kinds of processing on a track, and if you don't, the client isn't getting what they need. My approach is the opposite: I want to use the least amount of processing possible to achieve the best result.

I think a lot of younger engineers get overwhelmed because there are so many tools available now. But really, the best master is often just a little bit of EQ, a bit of limiting, and that's it. If you can achieve the result with those two tools, that's what you should be doing.

Over-processing is definitely an issue, but more importantly, you need to use your ears. Your best tool is always your ears.

Listen to the track. Don't just automatically run it through a bunch of gear without knowing if it will positively impact the mix.. Maybe it doesn't need all that processing—sometimes it needs almost nothing.

I have a track from [mixer] Michael Brauer that I'm working on today, and 95% of the time when he sends me something, I listen to it and send it back because it's already that good.

That's the hardest part of mastering: knowing when to leave it alone. If a track is great, just listen, do your job, and focus on quality control. The QC process is the most important part of mastering. If the song is good, your job is to make sure it's ready for production. I know that seems boring to some people, but that's what the song needs.

If the song is great, you don't need to put your fingerprint on it. There will be plenty of opportunities to do that, but when a track is already good, know when to back away.

18

RYAN SCHWABE

Ryan Schwabe is a two-time Grammy nominated, platinum certified mixing and mastering engineer. In 2020, he was nominated for a Grammy for Best Dance/Electronic Album for Baauer's "Planet's Mad," and in 2023 he was nominated for Best Sound Engineered Album, Non-Classical for mixing and mastering Baynk's "Adolescence."

Ryan is a former professor of recording arts and music production at Drexel University's Music Industry Program. He's also owner of the music production and engineering company Xcoustic Sound, the music technology company Schwabe Digital, and co-owner of the digital record label Rare MP3s.

And he's also the developer of the Gold Clip, Orange Clip and HiFAL mixing and mastering plugins that he talks about in the interview.

Bobby O: Tell me how you got into mastering. How did you learn it?
Ryan Schwabe: My experience in mastering was out of necessity. I was a young producer, engineer, and music creator and I was fascinated by the concept of doing the entire flow of music production myself. Mastering was obviously the last stage of that.

With that said, the first record that I put out that I produced, engineered and mixed, I hired somebody else to master it because I knew it was something that I didn't have a handle on yet. I certainly didn't want to print 1,000 CDs with a mistake that I wasn't aware of, so I just by handed mastering off because the manufacturing process was so expensive.

Nowadays the manufacturing process is nearly zero with streaming and the risk at the mastering stage is much lower, so many more people are taking it on themselves and more products are being marketed towards that DIY mastering space.

Didn't you get a request to master someone's project first?
I did. A friend asked me to master his rap album and lets just say I did my best at the time [laughs]. My career in mastering started in around 2006, but didn't become serious until 2010. I think Spotify came to the U.S. in 2008, so a lot of my work was purely based on digital distribution.

Early on, I wrote about loudness and digital distribution. I wrote about the structures of the way that digital distribution systems work between play listing and album mode and all of that. TapeOP reposted it and the internet herd came after me.

I think people started to look to me to master their projects because I just had a base understanding of those systems, in addition to what's culturally appropriate for different genres of music. There are two sides to mastering: There's the art of the sonics and the technical bridge to manufacturing whether that's CDs, streaming or vinyl.

How long did it take you before you felt that you were good at it?
Back when I started I was doing maybe 50 to 100 projects a year but I grew every year to where I think now I'm doing about 500-700 projects. Thats singles, EPs or LPs. It took me maybe a hundred per year for seven years before I felt a level of expertise. It's not really like a time spent thing though, I think it's an iteration thing.

I never got the wisdom that comes with sitting in on mastering sessions. I did as a client a few times, and that's largely the only time that you really do get to sit in on mastering. Almost all of the top mastering engineers have second and third engineers, but they rarely have interns.

My career was very much autodidactic. I was learning and iterating as I went along and making a few mistakes as I could. I think the biggest mistake I made was when I did an album for vinyl that was printed a thousand times, but the spacing was not technically correct, and so it had to be recalled.

That's an environmental disaster and a financial disaster, but every mistake you make is a mistake that you never make again if you're committed to lifelong learning.

Can you hear the final product in your head before you start?
Yes I can. I can imagine the power and detail I want to pull out of it. I can figure out where I want it to go. The direction may be aided by the reference mix that they give me or more often by my own internal reference points.

But yes, I can hear its final presentation before I even start. If you can't imagine it before you start you might just make it different, but not necessarily *better*.

You can make something different or you can make something better.
Yeah, and the delta between those two things rests in your cultural awareness of what's appropriate for that genre and your technical ability.

A lot of times when I was teaching mixing and mastering, my students would just make a bunch of things different and none of it was necessarily more compelling or *better*. They were doing things that weren't always appropriate for that genre or for the song, or took away from the main theme of the production.

In mastering, we're generally dealing with the last 5 or 10 percent of a record, so you really have to ask if that 5 percent is more appropriate for that song. If not, you might be better doing nothing at all.

Is there a typical problem that you see in mixes that you get to master?

Yes, it's probably three things: compression, low-end balance and overall clarity and detail of a mix. Engineers who aren't very experienced tend to misuse or overuse compression. The ability to hear attack and release isn't fully developed when someone first starts making records.

Dynamic effects are difficult to identify because you're listening to small changes in amplitude that only last milliseconds on the front edge of a sound. Or how those bits of gain reduction swell back to 0 and reinforce the rhythm of the music.

A lot of engineers early in their career can't hear those nuances, they just assume that everything needs to be compressed. Even if you are good at compression and can hear those details you might still overuse it.

Secondly, sometimes the low end is the wrong size in relation to the rest of the mix. Or, the relationship of the first three octaves of the low end is unevenly distributed. If the first three octaves of the low end is improperly distributed between the bass and kick and whatever else is living down there, the mix can sound heavy or sluggish.

Another glaring issue is when mixes sound smeary and *phasey*, and are not precise, deliberate, and intentional. This happens when a mix is too over-processed. Maybe every insert is used or there is unnecessarily complex routing. Or you just spent too much time on it and overdid it.

When you do too much, things tend to fall apart. But when people send me things that are over-processed, I don't say, "Oh, you did way too much. Start over." That's not a note that you want to get from a mastering engineer because even if it is true, it's arrogant and disrespectful. I try to be helpful and point out more obvious issues.

I give them notes on broader things; and it's usually level-based. For instance, maybe the vocal is hyper-compressed and *essy*, or the over all mix is missing energy above 8kHz, or an instrument is too loud or resonant in a specific frequency range.

Those types of notes are far more helpful and impactful to the quality of the final outcome of a mix at the mastering stage than a deep dive into the technical details of the mix.

Is there a level that you like to get mixes at?

I usually request to leave everything on the mix buss minus the final limiter. Compression, EQ, saturation - any processing on the mix buss is fine, leave it on, just bypass the final limiter. Sometimes when you take the limiter off, mixes completely fall apart. All of a sudden the mix is bass heavy and sounds like a foreign object to the artist.

That's not good. In that case I request they give send me low level pre-mastering, and a loud reference version. Then I can choose what to use and reference the approved version for loudness / tone matching.

I don't care about peak levels as long as they're below zero and not clipped. I don't necessarily care about the RMS value because I'm going to adjust the linear gain of the mixes before I start mastering anyhow. I'm going to adjust the level of the individual songs to set up an arc to the album.

When you're working with some of the more seasoned engineers, they will not send out low level premasters. They will only send out loud versions and that's totally fine too. I don't fight for unlimited premasters, but I personally think it will lead to a better master. I pick those battles based on my relationship to the client or engineer at that point.

However, it's their production, and if they feel strongly about it they might not want much to change in the final version. If I hear obvious things like intermodulation distortion I'll ask for a low level pre-master and screenshot of their limiter so I can recreate it. I'll recreate the exact same thing, but I'll be able to do a little more with dynamic processing.

Sometimes people don't want that either and that's fine. I can work with it. But I'll be like, "Okay, then this will most likely have to be a digital master [using only a digital signal path]," because I find that often times using hardware does not work too well with hyper limited mixes.

Do you use the analog or digital signal path more?
I'm always using the digital signal path and most of the time I am using both. But I am not a 'hardware is always better' type person. Well, I make plugins…[laughs]. I find that analog will add a level of saturation that makes the mix more mid-forward to some degree.

That rounding and a pulling the mid-range forward can be beautiful and natural sounding. And other times the precision of a digital master is far superior. It truly depends on genre, production and mix. I have a few plugin development teams so I am always experimenting with new custom processors in my workflow.

Do you have a digital signal path that changes often, or are you always using basically the same plugins?
I'm changing it like every two weeks [laughs]. I have a series of maybe 16 plugins that are all on bypass except for like Gold Clip [Ryan's own signature plugin] and a final limiter. And the components in Gold Clip are pretty much in bypass as well. All of the plugins in the chain serve some sonic problem that I'm trying to solve.

As I mentioned I also have custom Schwabe Digital high frequency tools, tube emulations and multi-band effects that I am building and experimenting with.

The series of mastering plugins all do something different. Schwabe Digital HiFAL controls the mid-range and tops. Gold Clip does upward compression without attack and release. So, each one of the processors has a specific sonic goal, but they're all off by default and everything is flat by default.

The first thing I do when I get a group of mixes is set the arc of the album. Then, I set up those linear gain modified pre-masters through the final loudness processing. Its basically, Gold Clip, HiFAL and a final limiter of choice and thats it to start.

Tell me about the plugins that you lean on the most.

In terms of loudness, iZotope's Ozone Maximizer seems to offer the most range I can get out of a single limiter plugin. It allows me to shape how I want the limiter to work. If I want to enhance transients, I can.

If I want to smooth the mids and deepen the subs, I can. If I want things to sound flatter, softer, and less exciting, there's also a setting for that. Ozone lets me do a lot of micro dynamic shaping in terms of how it handles loudness, so it's a primary tool for me.

Gold Clip is something I consistently use as well. You can use it in so many different ways — it's quite a nuanced device. If it's a hyper-limited mix I may just use box tone or Alchemy 2. Sometimes I do not use Gold Clip as a clipper, but as an upward compressor with the Gold processing.

For instance, if the peak information is around -2dBFS, I set Gold Clip ceiling to match that. The clipping meter shows 0dB of clipping, but I'm doing upward compression with Gold processing. It lifts the overall loudness of the mix without changing the peak information.

So, for example, if I increase the loudness by 1dB using Gold, I can reduce the final limiter by 1dB so it's not working as hard. Or, just click on unity and Gold Clip will drop all the peaks down by the same amount of gold I added . That maintains loudness out of gold clip, but the peaks are lower and the limiter is doing less.

I also really love the Metric Halo Sontec EQ. I use that a lot because it has a subtle beauty to it. I use the Kirchhoff EQ from Three-Body Technology and Plugin Alliance, which is a really clean parametric EQ. You can really get in and shape the mix or deal with resonances.

As I mentioned earlier, Schwabe Digital HiFAL is something I use on every record to some degree. It's very clean and versatile. It is a high frequency acceleration limiter, which is basically a de-esser with a variable attack and release dependent on the amount of gain reduction it's doing on the high frequencies. It also has a parallel processor built into it that adds static high frequencies to the mix. That has a big impact on the way top end is sitting in my masters.

I use the Michelangelo from Toneworks depending on the style of music. It's almost too colorful of an EQ a lot of the time. The plugin is quite versatile and you can actually adjust the amount of tube compression that it does.

I sometimes use the MAAT FiDef JENtwo, which is a very weird plugin that kind of just works. It's basically a noise generator triggered by amplitude and frequency of the input.

As the signal gets brighter or louder, it changes the noise that rides underneath. It's supposed to emulate the sound of analog, and the way it does that is by generating noise. Usually it's foolish to add noise, but this does something that's very unique and impressive. It's a very subtle psychoacoustic type of effect, but I usually prefer it on to off.

I do some widening with a Leapwing Stage Two, which is very phase coherent and smooth sounding and allows me to expand the stereo field in a way that's almost like an EQ that brings the midrange forward.

I usually use two limiters. The first Limiter is getting the final loudness, and then I'm doing some processing in between, and then a last limiter catches any peaks, I'll have a really clean shelving EQ after the first limiter that pulls down 3kHz and up by maybe a half a dB.

Limiters can tend to make things brighter as they get loud so that is a process that mitigates the added brightness that comes with limiting. That may sound like a lot, but I try to keep it minimal.

Let's talk about your Gold Clip plugin.

That came from working in analog as a mastering engineer. I used the famed Lavry Gold converters that everyone said clipped really well. People loved the loudness you could get from it, but to me it was not the clipping that people loved, that was an oversight. It was the sound of their soft saturation, which is this nonlinear gain process coded into the device post capture.

That was the sound of a lot of records when those devices first came out, especially when records started to get a lot of criticism for being loud. The Gold converters were largely responsible for the loudness of the 2000s. I used it quite a bit on the lower, more subtle setting of 3dB. I loved it so much I would print individual tracks of mixes I was doing.

So, I started to wonder if I could adapt that interesting gain manipulation and use it in a more variable way as a plugin. The hardware only had 3dB, 6dB, or Off settings, but what if I wanted to use like 0.4dB or 4.5dB? So I went down a rabbit hole and made something completely new and what I thought of as innovative.

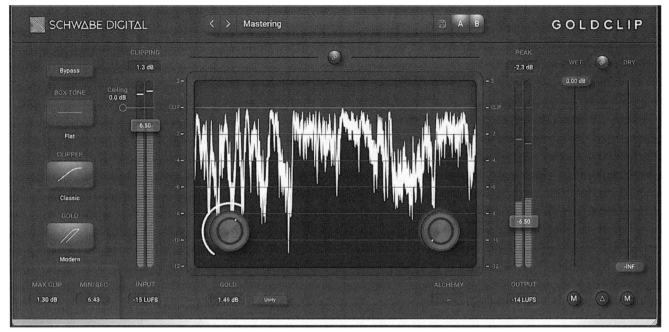

Figure 18.1: Schwabe Digital Gold Clip plugin

I use it first in the chain in my mastering because it has really nice gain management with linked inputs and outputs and a variable ceiling, so it lets you adjust the gain structure in multiple different ways.

That was my first entrée into plugin development and it did well. It was a blessing and it's on so many hit records right now. I built it as a mastering device but I noticed that a lot of people use it in mixing as well. I was just talking to Stuart White and he said he used it on the new Beyonce *Cowboy Carter* record.

Tell me about your monitors.
I'm using Kii Audio Three speakers. They're digitally connected to my computer through an AES output from my DAW, going into a Trinnov processor which is doing some room correction, and then into the AES input on the Kii's. The digital to analog conversion is happening in the Kii's.

There's zero noise in the system. If I hear noise coming out of the speakers I know it's in the mix and not in a speaker switcher or cabling or anything like that. It's life changing. I remember I had everything set up in analog for six or seven months before I figured out how to set up my system digitally, and it was just a night and day difference.

I also have two pair of headphones. I have the Audeze LCD-5s and the Dan Clark Stealth headphones, and those are both being fed from a Chord headphone amp.

How much do you care about LUFS levels?
I care about the way that the music lands and feels coming out of speakers and not so much about a numeric value. I'll usually start with the single or the most impactful song, and get it as loud as I can before it starts to fall apart, and then bring it back a dB or so. If it's really loud and it still feels engaging and it's appropriate for the genre, then that's where it lives.

Sometimes, it's artist dependent. Some artists are like, "I want this to be really natural and open. I don't want any unnecessary intensity to my music." Then that's what's appropriate for that project and the LUFS are naturally lower, but generally, I think a nice average level is between -7 or -10 LUFS.

I want to push things to what is appropriate for the genre and production. That's accomplished through an awareness of what works for that style of production.

Sometimes that means it's loud and sometimes it's not. Sometimes I'm delivering -4LUFS on a hyperpop style record. Sometimes I'm delivering -12LUFS masters on a folk record, it really depends.

What are you delivering generally?
I master at 96kHz. I'll upsample and work at that rate. That's the mother file, and then I deliver downsampled versions from there. I'll deliver stereo interleaved 24/96kHz bit Wav files, and then I'll also deliver it as 24bit/48kHz.

Most digital distribution systems are now accepting 24/96kHz, although some are still only accepting 16/44.1kHz. I only deliver 16 bit/44.1 and MP3s if people specifically ask. Since Apple went toward Atmos, I haven't gotten a lot of requests for Apple digital master lately, either.

I also deliver alternate versions of songs. Modern deliverables can be quite extensive. They may need the main version, TV, clean, instrumental, and an a cappella version. There might also be a clean acappella, and clean TV.

So that's sometimes six or seven alternate versions per track. I did the Fridayy record that had six or seven alternates per track and it was a 12 track album, so we're talking 90 deliverables of alternates and mains for one album, not to mention the Atmos mixes and all the stems the labels want.

Are you doing a separate master for vinyl?
Yes, but whether I create a separate print depends on the style of music. Sometimes I'll deliver a 1-3dB down version for vinyl. Sometimes I'll just deliver the digital master because there's so much high frequency control in my masters from my HiFAL plugin.

HiFAL does the acceleration limiting that controls the high end, so it won't burn out the cutting head. It really depends on the production and style of mix, whether I decide if it needs a whole separate master or not, because oftentimes it does not.

I was doing an RJD2 record and I sent him the masters for the record. He was like, "These sound amazing," but he never told me he was doing a vinyl release as well.

Months later he called me up and said the vinyl press sounds amazing. He took my digital masters, which were relatively loud at like -9LUFS or -8LUFS program level and made the vinyl from them and it came back sounding amazing. It was loud. It was crisp. It was detailed. There was no distortion in it. It just sounded perfect. And I was like, "Hmmm, I've been making all these lower level masters for vinyl, thinking that was better, and this sounds like the best vinyl that I've done so far."

Take your wins when you get them, but that was like a little bit of a surprise to me. I think the outcome of vinyl depends on the style of production, because his stuff was very smooth and natural sounding, so it worked. If it was lightning bright pop record I might not have been so lucky.

There are so many mastering plugins these days, plus the online mastering that an artist can use. How do you feel about that?
I think it's cool. People sometimes send me their reference masters from LANDR or something they made themselves with DIY mastering tools. I do what I do without listening to it, and then just AB it real quick. I always win, so I haven't felt it's going to replace anything I do, but I think they are wonderful tools. Anything that helps people make music more easily and confidently, I'm for it.

One problem is that it doesn't give people a chance to learn and actually have the knowledge base that I've been graciously allowed to learn over time. The tools often make the decisions for you and you do not get to learn much about what is happening.

I could see that technology being a problem for younger master engineers, and I could see it eventually becoming a problem for younger mixing engineers, because the tools are becoming more sophisticated everyday.

It's letting people get results fast. In some ways I feel like the technology is making mastering less of an art and more of a technical process, which is unfortunate. But I don't feel a threat from it.

What's your policy on revisions?
I have no policy on revisions. You can revise however much you want, but I know that when I send you the first version, you're going to be like, "Oh my God, this is amazing." I don't remember the last time I got more than a revision or two and 90 percent of the time it's version one and it's done.

That's because I know exactly what's its supposed to sound like, and know the best solutions for the particular issues that may be presented, which is a big part of being a mastering engineer.

I think it's that breadth of experience that provides for a more efficient path to the final product. You may eventually get there with somebody with not as much experience, but it may take a lot more versions.

When you finish you might wonder, "Was that the path that we really wanted? Do I feel more confident about this being the end?" The final mastering process should be a big high five on V1 so people get excited about pushing their art to the world. Thats my goal every time.

What's the biggest difference between a really great mastering engineer and one that's just competent?
Great mastering engineer is someone who has equal parts deep technical knowledge and deep cultural knowledge. Their cultural knowledge gives them the right instincts and their technical knowledge lets them communicate those instincts.

I believe that somebody that doesn't have experience in a genre, or hasn't seen the whole spectrum of sonic issues presented to them in a mastering process, is more prone to make mistakes or overdo it. Or they identify things as problems that are not problems. Thats when you land on masters that you do not like.

I've mixed records that were mastered by legends and I was like, "damn, I don't like this…" but I would say the rate of a bad outcome with a legend to the rate of a bad outcome with somebody just starting out is like 150X.

So that's the value of what you get with a great mastering engineer. It's the 20 years of experience that they have making decisions about impactful music. They arrive at better solutions faster. When an artist is at the very end of making their record they deserve to get masters that make them celebrate their work, not question it.

19

IAN SHEPHERD

Ian Shepherd is a highly respected UK-based mastering engineer with over 30 years of experience. Starting his career at Sound Recording Technology (SRT), Ian spent 15 years honing his skills and mastering for major artists, labels, and orchestras before striking out on his own.

In addition to his mastering work, Ian is a prominent educator and advocate for high-quality audio. He founded *Dynamic Range Day*, an annual event aimed at raising awareness about the importance of preserving dynamics in music, countering the trend toward excessive loudness. His educational platform, *Production Advice*, offers courses and one-on-one coaching for those looking to improve their mixing and mastering skills, and he hosts *The Mastering Show* podcast.

Ian is also a plugin developer, co-creating innovative tools like Perception AB and Dynameter as well as the Loudness Penalty website, which help engineers achieve the best possible balance between loudness and dynamics when mastering.

Bobby O: How long did it take you before you felt that you were confident at mastering?
Ian Shepherd: Any day now? (laughs) It probably took a year or two before I really felt like I had a solid handle on mastering. I started with copying tapes in my job at SRT, then moving to doing sessions on my own, and eventually I began working on attended sessions with clients.

Having the client in the room with you was a whole different experience and I definitely had some insecurities about it at first, but the sessions went well, and the clients were happy.

The funny thing is, I recently found a couple of the first projects I ever worked on through eBay. I put them in the CD player, half expecting them to sound terrible, but they sounded pretty good! I was really pleased with how they turned out.

I was fortunate to have mentors and that gave me a safety net, which made the whole process feel fairly comfortable.

That said, you're always learning. Now I look back and think, "You knew nothing back then!" But at the time, I got to a point where I felt confident in my abilities and was doing good work, and that's the important thing.

Can you hear the finished product in your head?

It's funny because when I first started at SRT, copying tapes was really frustrating. I'd just be sitting there listening for faults in the transfer, but I always wanted to fix things that didn't sound as good as I thought they could have.

It wasn't about having a grand vision of how it should sound; it was more about recognizing when something wasn't there yet. I'd think, "Okay, this doesn't sound right to me—what can I do to make it better?" Then I'd follow that impulse as far as I could to get the track sounding as good as possible. I still do that today.

Is there a type of music that you find more difficult or easier than everything else?

In general, I'd say the more dense the genre, the harder it is to master. You're constantly balancing fidelity with loudness, distortion, and aggression. It's a tough juggling act.

That makes classic rock especially one of the most difficult genres because it has such a well-established sound that everyone expects.

We all have this imprint in our heads, probably from listening to FM radio in the '80s or whatever era. There's so much going on in the arrangements—they're very full with live instruments, which means you've got imperfect drum sounds, natural guitar tones, and so on.

There's a really delicate balancing act involved. You need just the right amount of top end, just the right amount of midrange, and you have to separate all those elements while maintaining that energy. It has to feel energetic and intense, but you don't want to squash it too much or you'll take the life out of it. That's what makes it so challenging.

Another genre I personally struggle with is steel band music! Even though I love hearing it live, for some reason, it just doesn't seem to record well. Maybe it's because the sounds of the instruments are so full of super-high harmonics, and they all seem to leap out in an uncontrolled way. It's always a challenge to capture that properly in a recording.

Do you find that there's a problem that keeps coming up in the mixes you get in?

These days, I think so many mixes are just too loud. We've all been talking about the "loudness wars" for what feels like forever, and most people know my thoughts on the subject!

Right now it's saturation and clipping that everyone's talking about. Two years ago, it was parallel compression. Every year, there's some new "secret technique" that pops up on the Internet, and then everybody jumps on it. You see tons of videos teaching people how to do it, and usually the focus is on getting things as loud as possible, but it often gets overused.

Back when I started, 25 or 30 years ago, people were making super clean, super dry, '80s-style home recordings that lacked saturation, warmth, grit, and density. As a result, that's what we added in the mastering process. Now, I feel like I'm doing the opposite. I'm trying to bring back some space, depth, and life into tracks that have been squashed and over-processed.

Fortunately, people tend to come to me knowing that I'm not a fan of doing overly loud masters, so I'm not the go-to person if someone wants everything cranked to the max. That means I get a higher proportion of clients who are already on board with the idea of balanced dynamics and well-managed loudness.

But yeah, that's the main issue for me—there's just too much emphasis on making everything loud for the sake of it. It doesn't leave me as much flexibility in the mastering, and it can take away from the overall quality and musicality of the sound.

I've found that when I suggest to clients to ease things back a bit and reduce the intensity in their mixes, they usually agree. And every time they do, we end up with a master we're both happier with.

Tell me how Dynamic Range Day came about.

The idea for Dynamic Range Day actually came from two blog posts I read when I started my own blog. One was from Derek Sivers, the founder of CD Baby, and the other was from marketing expert Seth Godin.

Derek talked about how someone has to go first—like the first person to dance at a party or the first to stand up for a standing ovation. Then Seth wrote a blog post about creating a holiday around something you're passionate about.

I thought, "Why not be the first to create a day to celebrate dynamics?" and that's how Dynamic Range Day was born. I kind of regret the name now because it's not strictly about dynamic range, but it stuck.

Originally, the idea was for everyone to type in ALL CAPS all day online, like shouting, which was funny but also really annoying! So I shifted to making it more positive over time.

It gained traction, and SSL reached out to sponsor a competition, donating an X-Desk as a prize. It became a yearly event, and while I'd love to leave the loudness debate behind, I still think it's necessary. Some people are tired of hearing about it, but when music I love comes out and I think, "If only it had a little more room to breathe," I just can't let it go.

What do you recommend for achieving balanced dynamics? I know it's not as simple as saying "10dB" or something like that.

Nowadays we measure in LUFS (Loudness Units relative to Full Scale), which is a bit confusing because there are three different types. The one used by streaming services is called Integrated LUFS, which is an overall value and doesn't really give enough information to make detailed judgements. I suggest people pay more attention to Short-Term LUFS, which is closer to what a VU meter would show in the old days.

My guideline is to keep the loudest sections of a track no louder than -10LUFS Short-Term, with True Peaks at -1dB to leave headroom for encoding to lossy formats like MP3. You want to ensure the loudest parts sound good first, then balance everything else musically. Ideally, the mix is already balanced, but if not, you can use a little gentle automation to adjust the quieter sections.

For albums with a lot of variety, you might end up with some tracks at -16LUFS or even lower, and that's fine. The loudest parts might be around -10, but as long as it feels right musically, I'm comfortable with that range.

If you want something to sound loud, especially in streaming, it needs to be at least -14LUFS for the platforms that don't turn quieter songs up. Most "loud" music will naturally end up louder than that anyway. I often find that -10LUFS for the loudest sections is a good place to be. It's loud enough to hold its own while preserving musicality.

The streaming services set their loudness normalization values based on a wide variety of material, so while people complain about the numbers, they came from analyzing what works best across a broad range of music.

What monitors are you using?
I use Bowers and Wilkins monitors, the same brand I had at SRT, though the ones I have now are smaller. I don't think they make this model anymore, but they work really well in my current room, and I'm really happy with them.

Are there plugins you find yourself using all the time?
I use my own... (laughs). When I left SRT it was the first time I didn't have access to the hardware I'd been used to and had to start experimenting with plugins. That's when I started using FabFilter, and I still think their plugins are excellent.

I also find TC Electronic's tools really quick and easy to get the musical results I'm after. I think it's partly because I used their hardware for so long and learned how to make the most of it. It's like the 1176 or LA-2A—sure, they have limitations, but if you know how to work with them, they're perfect for what you want to achieve. The TC plugins may not have quite as many features as more recent stuff but I use them on pretty much every job.

Do you have a typical signal path that you would use? And how many plugins would you say is average for you?
I'm 100% in the box these days because I have everything I need. In terms of workflow, efficiency, and project recall, working entirely in the box just makes so much sense. My chain is fully digital unless someone specifically requests something analog.

One of my favorite sayings is that mastering is simple, but that doesn't mean it's easy.

For me, that means the standard chain is EQ, compression, and limiting. If you have one good unit handling each of those stages, you can get great results on about 80% of the material. Sometimes you might want to use multiple stages of compression, swap out the limiter for a different sound, or use clipping along with limiting—but the core chain is usually the same: EQ, compression, and limiting.

Another common element in the chain is stereo processing. This can be as simple as adjusting the mid-side balance, or something more involved like using Leapwing StageOne for additional control.

Sometimes I might bring in something like the Michelangelo EQ plugin to add a bit of character. I also really like Wavesfactory Spectre, which is a multiband saturation plugin. For example, you can add saturation to just the low mids to bring out aggression in the guitars without actually changing the EQ, just by adding the harmonics and distortion you want. Occasionally I'll use tape emulation for a similar flavor.

I like working this way because you can introduce these effects at the end of the chain, but excluding metering, the chain is typically three main plugins: EQ, compression and limiter.

How did your own plugins come about?

The idea for Perception AB was to create a plugin that allowed loudness-matched bypass so you could hear both the pre- and post-mastering versions with fast, automatic loudness matching.

In mastering, we often check the level of the mastered version balanced against the original mix to make sure it actually sounds better and not just louder. Learning how to loudness-match is one of the key skills in mastering, and Perception AB was designed to make that process faster and easier.

I had always liked the original TT Dynamic Range meter plugin (no longer available, but reincarnated as the MAAT Digital DRMeter), which shows you the crest factor –which is not technically the same as dynamic range, despite the name.

I thought the TT meter was cool, but it would be even better if we could use LUFS rather than RMS, and track that information over time as a color-coded graph. That's where the idea for *Dynameter* came from. It's a great way to see the peak-to-loudness dynamics of a whole song at a single glance, or as it plays.

It was another one of those things where I thought, "This would be cool to have," and I was lucky enough to find a great partner in Meterplugs to help bring them to life.

What's your take on online mastering services or mastering plugins?

I think AI-based plugins are really cool, but it annoys me that they call themselves "mastering" plugins because they're not really mastering. It's more like automated EQ and dynamics balancing. It's like the wizard in Photoshop that automatically makes an image look "fine." A lot of the time it does a reasonable job, but sometimes it's horribly wrong, and other times it's just "take it or leave it."

Mastering is about having a conversation with a person. It's about me understanding the artist's intent and trying to help them realize it, to make their music the best it can be. It's also about understanding that some songs are meant to be quiet, others loud, some are meant to be angry even though they're quiet, or fun even though they're loud.

Mastering is about grasping the emotion behind the music, and AI tools just aren't there yet—and I'm not sure they ever will be.

That said, I do think people are using these tools in creative ways. I have a lot more respect for iZotope's approach, where instead of a black box where you put something in and it spits out a finished product, they let you see what the tool is doing and give you the option to adjust it. I've played with their Mastering Assistant—I don't use it professionally, but I've tested it.

Earlier this year, I did a lecture at Leeds Conservatoire where we compared three or four different AI algorithms with a master I'd done myself. When we listened to each track individually, they sounded "fine", but when we played track one and then moved to track two, the second track was way too loud, even though it was meant to be a quiet song. The algorithm hadn't figured that out. So, yeah, I think AI in mastering still has a long way to go.

With the Mastering Assistant, for example, I would probably turn off four out of the five modules it suggests and keep just one, because sometimes that's the one that does something useful. And I think that's a useful learning opportunity for people.

If you can't afford or don't want to go to a professional mastering engineer—which is always the best option—I think your time is better spent learning how to use a good limiter and really diving deep into an EQ. For me, that's 80–90% of the battle. If you can get the EQ curve right and choose the right balance of loudness and dynamics, that alone can transform a track.

I actually offer mastering courses, and one thing a lot of people tell me is, "I learned from your course that I don't want to be a mastering engineer, but it was so helpful for my mixing and recording." It gives them perspective on what's important at each stage, what they can deal with in mixing, and what can be addressed in mastering.

What are the files that you normally deliver?
I'm still delivering DDP files, but for different reasons than before. It used to be because they had built-in error correction and clients couldn't easily tamper with them. I'd give clients software to preview and burn CD refs from the file so they could check it, sign it off, and then send that same link to the plant. That way, I knew the plant would have exactly the same file I sent out, so it was great for quality control.

These days, most clients are asking for high-spec files. Even so, I still deliver a DDP first, even if they aren't having CDs pressed. I don't do this for single songs, just groups of songs. I like the fact that the DDP includes gaps, metadata, and everything else built-in. Ironically, what used to be the actual production master is now the "low-quality" option that clients are less interested in, and I supply the high-spec files once everything is approved.

Do you do Apple Digital Masters?
Yep, I'm Apple Digital Masters certified, though I don't get people asking for it that often. I offer it as part of my service. I know some places charge extra for it, but everything I do already follows their guidelines, so I just include it. If a client cares about it, I can say yes, I'm certified.

I also supply things like MP3 files and other formats if people ask for them, but I don't feel it's worth their while because the metadata we put into the files is usually ignored by aggregators. Clients will have to input all the metadata again when they resubmit anyway, so it feels redundant.

Plus, I think lossy formats like MP3 are temporary—the goalposts keep moving, and I don't see the point in optimizing heavily for them.

What about vinyl masters?
Again, I deliver files, but it depends on the plant. Some plants want Side A and Side B as single files, so I'll provide that if needed.

I have this concept I like to call "one master to rule them all," though I've seen others using that phrase recently so maybe someone else came up with it too. Basically, the same master I use for CD, streaming, and everything else is also the vinyl master—except for the vinyl I disable the final brick-wall limiter and drop the level a bit. The cutting engineers I work with prefer this because less limiting helps them get a cleaner cut.

That said, the masters I do aren't loud enough or heavily limited enough for it to be a huge issue. The only exception is when the limiter is contributing something important to the sound. Normally, the final limiter is just controlling occasional peaks, so it's fairly "invisible." If it's playing a bigger role in the sound, I might keep it in, even for the vinyl master.

What workstation are you using?
Steinberg's Wavelab.

What do you think is the hardest thing that you do in mastering?
It's when something sounds terrible, and there are two scenarios: either the client doesn't realize it sounds terrible, or they do. If the client doesn't know it sounds bad, it's easier, because they likely won't be too critical of what you've done. But the really hard situation is when the client knows it sounds bad, and they want you to fix it!

In that case, you're doing everything you can to improve it, and while they can hear and appreciate the improvement, they still want more. They keep asking for more, even when you've reached the limit of what's possible.

The challenge is knowing when to say, "I'm really sorry, but this is as good as it's going to get. If you want it to be better, you'll have to fix X, Y, and Z in the mix."—but they don't want to hear it. That's definitely the hardest situation!

20

HOWIE WEINBERG

Howie Weinberg has won 20 Grammy Awards and 76 Grammy nominations, four Tech Awards, two Juno awards, one Mercury Prize award, 200 plus Golden Platinum Records, an unbelievable 8,600 total credits that have accumulated over 91 billion streams!

He began his career working in the mailroom at Masterdisc in New York City, but soon became the apprentice of mastering legend Bob Ludwig. Within a few months, he was mastering tracks for hip hop stars like Curtis Blow, Run DMC, Grandmaster Flash, and Public Enemy.

Since then he has mastered projects for musical royalty like U2, Nirvana, Sheryl Crow, The Clash, Madonna, Alice Cooper, Aerosmith, John Mellencamp, Ozzy Osbourne, and the list goes on.

Bobby O: I know you have an interesting story about how you started in the business, so let's go there. Didn't you start from the bottom?
Howie Weinberg: Well, if you want to consider delivering packages and being the studio's first messenger, yeah. I started at Masterdisc in about 1980. I had no experience as an engineer, and they said, "We need somebody to deliver packages. We've been paying a Messenger service, so you could be our company's first messenger boy, and then after you're done working, you can come hang around the studios." So that's what I did after I delivered packages all over town.

A lot of my first duties as an engineer was to make tape copies for anything that was done on Mercury Records, Polygram or Polydor, so that was a lot of hours of work.

When did you get your first break to do your own mastering?
Well, that was another thing. The company had moved to a different location and they had an older engineer there. I don't know if they didn't like him or he wasn't working out, but one day they said, "We're firing him. You go in and work there."

And I had no idea what I was doing. My new studio had an antiquated system that they knew they had to get rid of, but they put me in there anyway and that was exactly when hip hop and rap music started coming out.

This guy called me and he goes, "I have this rap record for you to do." I didn't realize it was a Curtis Blow track called "Christmas Rapping," and it turned out to be one of the biggest selling rap records. I think it was the first gold 12 inch record.

I became known as the guy who did all those crazy hip hop records after that. I had so much work, but I wasn't really that happy with the equipment that I was using. I just couldn't get the sounds I wanted from that room. Bob (Ludwig) had a nice Neumann console with dual lathes and big monitors, so I made a deal with the studio that when he left for the day at six o'clock, I would use his gear.

That went on for maybe six months or a year, then we realized that the gear wasn't working out, so they built me another studio with the good equipment and I was off to the races. I would do four, five, maybe six albums a day. I started doing every record under the sun from the hip hop records to rock records to pop records. I can't even imagine how many records I did.

When did you go out on your own?
That was in 2009. There was a point where I was doing countless numbers of big hit records and all the clients were mine. The studio convinced me to stay there for longer and longer, and then after a while, it just wasn't worth my while. I mean, all the clients were mine and I was doing all the work, so I realized that I can just do this on my own without a problem.

At the time when I left in 2009, the whole vinyl thing had collapsed so you didn't need any disc cutting equipment, which was a really big financial problem when you started a mastering studio at one time. Plus there were so many things that could go wrong with it so there was a lot of maintenance involved.

We were digital by then so I didn't have to worry about any of that, and if I needed any lacquers cut I would send the files to Carl at True Tone and he would do all my cutting work. Then I could just set up on my own without worry about any of the cutting gear.

How long did it take before you felt really comfortable about the masters that you were cutting?
After maybe six months or a year I got a feel of what I could do to take a recording to the next level within the limitations of the format. I mean, the vinyl record itself is a limitation.

For instance, it doesn't sound as good from the middle of the record to the end, and if you want 10dB more bottom, it's going to overcut [which means it won't play on most record players]. You can't do something like put 10dB more bottom end on or make it super bright without causing chaos with the cutter head.

I mean, no matter how you slice it, vinyl records are like the Flintstones. You have a rock and you're cutting plastic with it. People think it sounds better, but I've never heard a vinyl record that sounded good to me. There was always something that went wrong. Either pops, ticks, overcuts, or you have to limit the bottom. You have to take everything and put it in the center just to make it fit on that Flintstones type record.

When we first started mastering for CD in the 80s. That's when I realized that I could do all the stuff I wanted to do and make everything really push the limit. I tried to do that with vinyl records a lot, and it kind of bit me in the ass sometimes. You'd have to recut things over again if there was an overcut or it was smearing the sound or something like that.

What do you do now when you have to deliver a master for vinyl? Do you make a different master from your CD master?

No, that's a fallacy. You master for perfect sound and that's it. You don't try to master for a format or it'll bite you in the ass in the end. And that's how I work. The only thing is there's the same problem I had 40 years ago - there's only a certain amount of bass that you can put on.

Vinyl has never had a good sound. They remind me of bad sounding CD. I think people like it because of its big covers, but for perfect sound, no, not happening.

How much has your setup changed from when you started to where it is today?

Not as much as you might think. I still have those big Sontec EQs. I have a set of these old NTP compressors like I had in the eighties. I don't use that old Neumann stuff anymore. I have some very modern SPL equipment that has jacked up power supplies. The only thing I would say that's better is that the analog chain has gotten a bit cleaner and the clocking is really as good as it can possibly be.

I used to think digital clocking was all the same. It's just ones and zeros. But when I heard first heard those Antelope Atomic Clocks, it was like night and day. It was like this one sounds beautiful, and the other one sounds mushy, you know?

As for digital hardware, I don't really use it much anymore. Honestly, I don't think it's any better. It may look impressive to clients, but when I work on a project, I now master everything using digital plugins and an entirely digital setup. At the same time, I also master it using a full analog setup, which includes my SPL analog front-end, my Sontec EQs, SPL compressors, NTP compressors—basically a full analog chain alongside the full digital one for every project.

I reached a point where I was getting too many recall requests. If I used analog, clients would say, "I want it to sound more like what we gave you." If I used digital, they'd ask for more color or warmth. So now, I just do everything twice—once in digital and once in analog. It takes a little more time upfront, but in the end, it saves time by reducing recalls.

This way, the client can choose between the two options—one that's very close to what they initially gave me (the digital version) and one that has more analog color. So whatever sound they're looking for, they've got it. My workflow is much smoother now, and I don't get as many redo requests as I used to.

When you get mixes in, do you see the same mix problems over and over?

I'm not that judgmental like I used to be. I'm not a producer. I'm not a mixing guy, so whatever somebody gives me, if this is their best, I'll work with it.

I work on a lot of big records and with a lot of well-known engineers that provide loud reference mixes to their clients that is going through a digital limiter and getting really loud. They're giving me that reference mix that's super loud, and then they give me the normal mix that's uncompressed.

A lot of times I get a mix and I think, "Wow, their reference mix is so loud that I can't..." The point is, you want to give them something as good as, or better than, what they've already done. I don't recommend pushing it that far, but sometimes I have to. There are times when I have to make things louder than I'd prefer, just to match the level of their reference mix.

And that's fine, but it's not always the case. A lot of producers I know have reference mixes that I can surpass pretty quickly, and once you do that, you're ahead of the game.

What do you tell someone that asks what would help you work the best? Do you tell them to take the limiter off?

Well, yeah, just give me both. I want to know what the client's listening to. It goes back to sending reference mixes out to the artists, because they want to hear everything louder. Generally speaking, I can always beat a reference mix.

What's the most fun for you?
I really enjoy working on good projects with great artists. It makes the work fun for me. These days, I can probably handle around 100 to 150 albums a year, maybe even 200. I usually work on about 200 songs a month, sometimes even more.

What's great is that I get paid to play with people's music. I have a studio in my house, and what's fun for me is creating good music while making a living doing something I'm really good at.

I work with a wide range of artists—new, upcoming indie artists, and major artists alike. What I'm really good at is collaborating with all kinds of people. I don't have an ego about it. I'll work with the band down the block or a big platinum artist, sometimes even at the same time.

I also do a lot of work for music composed for TV shows, films, and video games, including companies like EA. Everyone has different budgets, so the first thing is figuring out what I can provide based on that. Every project is unique, and every client is different, but that's fine—it's all part of the job. The project turns out great, and everyone's happy.

What's one thing that you wish people knew about what you do? Is there one thing that they just don't get?
I don't really want to boast, but I can complete a project in 30 or 40 minutes that might take someone else two or three hours. I've worked on so many projects in my life that it's almost automatic for me at this point. I can listen to a track and immediately know where it's supposed to end up. I don't need to spend a lot of time figuring things out.

That being said, there are still some projects where you have to really focus and zero in on the details. Sometimes you get a mix that's just perfect, and all you need to do is put your spin on it, and everything turns out great. Other times, you get challenging mixes that require a bit more effort to get just right.

Is there a genre that you find more difficult than others?
A lot of the hip hop stuff tends to be overloaded and louder, but with pop rock, I've done so many of those that I just know what's going to sound good. I've got great monitors in my studio, ranging from big monitors to medium, small, and even tiny bookshelf speakers. I typically rely on two sets of monitors that I trust: the big ones, which are more fun, and the smaller ones, which are more practical. You just develop an instinct for it.

It's hard to explain—you put something on and think, "Maybe it could be a little brighter," or "This is really good, I don't want to mess with it too much." It's all about knowing when to adjust and when to leave it alone. I could have 500 buttons and knobs, but I only really use four or six, because they're the ones that matter.

When I take on a project, I also ask the client if they have any specific direction or preferences. Sometimes they do, and sometimes they don't. Either way, I aim to deliver both a solid digital and analog version so they can choose what works best for them. This approach helps me streamline my workflow and reduces the number of recalls I get.

What monitors are you using? Are they the same as what you started on?
Yeah, since I've been here, I got the big PMCs, the large ones. The room is tuned to them, and they sound great. I like them because you can play things loud, and I love playing loud music.

My new favorites, though, are these monitors from Keith Klawitter. They're fantastic speakers—forward-sounding, not dark, but bright and energetic. I love them, and clients really like them too. They've got a unique sound, and they're powered, so it's just plug-and-play.

I also have an old set of Dynaudio BM15s, the Genelec 1031As (which are kind of classics), some little Advent bookshelf speakers, and some headphones. But for my usual workflow, I either go with the big PMCs, the KD monitors, or sometimes the Genelecs or KRKs. There's no magic speaker in this world that's perfect for everything.

Tell me about your plugin.
Well, that plugin [the Howie Weinberg Mastering Console] is great, but there's a new one coming out soon. The old one is from Acustica Audio, and it's a pretty over-the-top digital piece. It's really good—it got me into a place where I could use one plugin to emulate all of my SPL gear, like the compressors and converters, and the whole setup.

But you need a powerful computer to run this plugin properly. If you don't have the right setup, it's just not going to work. I know some people who have it, and they absolutely love it. They couldn't believe how great it sounds. We spent a lot of time perfecting it.

My next plugin, though, is going to be much simpler—not as many knobs and buttons. It's in the works now, but nothing's finalized yet. It'll definitely be easier to use.

What's the best piece of advice that maybe you learned along the way or somebody imparted to you?
My main piece of advice is to always make the client happy, no matter what. Even if they're asking for something you don't think is the right choice, stay open-minded.

I remember a few years ago, I used to see Bob Ludwig working on a lot of records, some of which were huge, while others weren't as impressive. I asked him how he dealt with it, and he said, "For me, it's more about making the client happy." No matter what, whether it's an A-list project or a B-list one, the goal is the same: make the client feel good about the process and the final product.

Even if it's not your ideal project, if they leave thinking, "Yeah, this guy's great," they're going to want to work with you again. That's what counts in the long run.

21

BOB LUDWIG - GATEWAY MASTERING

Bob Ludwig is considered one of the giants in the mastering business, with 13 Grammy awards and 22 other nominations coming from his more than 3,000 credits. Among his massive credit list include legendary records from Queen, U2, Sting, the Police, Janet Jackson, Mariah Carey, Guns N' Roses, Rush, MötleyCrüe, Megadeth, Metallica, David Bowie, Paul McCartney, Bruce Springsteen, the Bee Gees, Madonna, Elton John, and Daft Punk.

Bob recently closed his Gateway Mastering and retired, but his thoughts on mastering are well worth reading, since it shows that the basics still haven't changed. He also provides a bit of history about "the loudness wars" that the industry suffered through, and how it came about.

What do you think is the difference between someone who's just merely competent and someone who's really great as a mastering engineer?
I always say that the secret of being a great mastering engineer is being able to hear a raw mix and then in your mind hear what it could sound like, and then knowing what knobs to move to make it sound that way.

You know where you're going right from the beginning then, right?
Pretty much. It's a little bit like the Bob Clearmountain school, where after 45 minutes of mixing he's practically there and then spends most of the rest of the day just fine-tuning that last 10 percent. I think I can get 90 percent of the way there sometimes in a couple of minutes, and just keep hanging with it and fine-tuning it from there.

It comes very, very fast to me when I hear something, and I immediately can tell what I think it should sound like. The frustration is, sometimes you get what I call a "pristine piece of crud," because it's a bad mix and anything you do to it will make it worse in some other way. Ninety-nine percent of the time, I hear something and I can figure out what it needs, and fortunately I know what all my gear does well enough to make it happen.

The loudness wars... Where did that come from?
I think it came from the invention of digital-domain compressors. When digital first came out, people knew that every time the overload light went red, you were clipping, and that hasn't changed.

We were all afraid of the over levels, so people started inventing these digital-domain compressors where you could just start cranking the level up.

Because it was in the digital domain, you could look ahead in the circuit and have a theoretical zero attack time or even have a negative attack time if you wanted to. It was able to do things that you couldn't do with any piece of analog gear. It could give you that kind of an apparent level increase without audibly destroying the music, up to a point. And of course, once they achieved that, then people started pushing it as far as it would go.

I'm totally convinced that over-compression destroys the longevity of a piece. When someone's insisting on hot levels where it's not really appropriate, I find I can barely make it through the mastering session.

Another thing that contributed to it was the fact that in Nashville, the top 200 Country stations got serviced with a special CD every week that had the different label's new singles on it.

When they started doing that, the A&R people would go, "Well, how come my record isn't as loud as this guy's record?" And that led to level wars, where everyone wanted their song to be the hottest one on the compilation.

I suppose that's well and good when it's a single for radio, but when you give that treatment to an entire album's worth of material, it's just exhausting. It's a very unnatural situation. Never in the history of mankind have we listened to such compressed music as we listen to now.

Tell me about your monitors.
I used to have Duntech Sovereign 2001 monitors when I worked at Masterdisk, and when I started Gateway, I got another pair of Duntechs with a new pair of Cello Performance Mark II amplifiers. These are the amps that will put out like 6,000-watt peaks. One never listens that loudly, but when you listen, it sounds as though there's an unlimited source of power attached to the speakers. You're never straining the amp, ever.

Every client that comes in, once they tune in to what they're listening to, starts commenting on how they're hearing things in their mixes that they never heard before, even sometimes after working weeks on them. It's great for mastering because they're just so accurate that there's never much doubt as to what's really on the mix.

One reason I've always tried to get the very best speaker I can is I've found that when something sounds really right on an accurate speaker, it tends to sound right on a wide variety of speakers. I've never been a big fan of trying to get things to sound right only on NS10Ms.

Do you listen only with that one set of monitors, or do you listen to nearfields?
Primarily just the big ones, because they tell you everything, but I do have a set of NS10Ms and some ProAcs and stuff like that. Lower-resolution nearfields have their place.

Do you think that having experience cutting lacquers helps you now in the digital domain?
It does. I'm certainly more concerned about compatibility issues than a lot of the mixers are, especially as more people are getting into synthetic ways of generating outside-of-the-speaker sound. Some people just get into this and don't realize that their piano solo is gone if played back in mono.

People do still listen in mono, but some artists just don't seem to be bothered by the lack of compatibility. Nevertheless, I'm probably more hypersensitive to sibilance problems than I would otherwise be if I hadn't cut a lot of disks.

Does that mean you still listen in mono a lot?
I certainly check in mono. We have correlation meters on our consoles. In my room, if you're sitting in the sweet spot and flip the phase on one of the speakers, the entire bass goes away. It's almost as if you were doing it electronically, so you can hear any phase problems instantly. Plus, I have the ability to monitor L minus R, as well as to hear the difference channel if I need to.

What workstation do you use?
We've been using the Pyramix workstation, and we have the Horus convertor that will do everything up to 384kHz PCM or DSD or DXD [Digital eXtreme Definition]. It's a really good system.

We play off of Pro Tools and we also have a Nuendo system that's part of an AudioCube system that supports a couple of plugins that are only on the AudioCube.

Does that mean you're going from the digital domain to analog and then back to digital?
It completely depends on the project, but normally, yes. And of course we still get tape. That Daft Punk record that I got three Grammys for came in on five different formats, including DSD and PCM, and the tape won out for that particular case.

If it's something like a mix from Bob Clearmountain where the mix is almost perfect to begin with and you don't want to do too much to it, then it works just to stay in the digital domain.

What's the hardest thing that you have to do? Is there a certain type of music or project that's particularly difficult?
I think the most difficult thing is when the artist is going through the period where they just can't let go of the project.

You get into the psychological thing where in the same sentence they say, "I want you to make the voice more predominant, but make sure it doesn't stick out." Just contradictory things like that. They'll say, "This mix is too bright," and then you'll dull it up like half a dB and they say, "Oh, it doesn't have any air anymore." It's that kind of thing.

Do you have a specific approach to mastering?
To me, music is a very sacred thing. I believe that music has the power to heal people. A lot of the music that I work on, even some of the heavy-metal stuff, is healing some 13-year-old kid's angst and making him feel better, no matter what his parents might think about it, so I treat all music very seriously.

I love all kinds of music. I master everything from pop and some jazz to classical and even avant-garde. I used to be principle trumpet player in the Utica, New York Symphony Orchestra, so I always put myself in the artist's shoes and ask myself, "What if this were my record? What would I do with it?"

That's why I try to get some input from the artist. If they're not here, at least I try to get them on the phone and just talk about what things they like. I just take it all very seriously.

22

DOUG SAX - THE MASTERING LAB

If ever there was a title of "Godfather of Mastering," Doug Sax has truly earned it, as evidenced by the extremely high regard in which the industry holds him. One of the first independent mastering engineers, Doug literally defined the art when he opened his world-famous Mastering Lab in Hollywood in 1967.

Doug recently passed away, but his magic remains a big part of the albums he worked on for major diverse talents such as The Who, Pink Floyd, The Rolling Stones, the Eagles, Diana Krall, Kenny Rogers, Barbra Streisand, Neil Diamond, Earth, Wind & Fire, Rod Stewart, Jackson Browne, and many, many more.

Although somewhat dated, this edited interview is included just to show that even though we have new powerful mastering tools available, the actual mastering process and philosophy hasn't changed much in over fifty years.

Do you have a philosophy about mastering?
Yes. If it needs nothing, don't do anything. I think that you're not doing a service by adding something it doesn't need. I don't make the stew, I season it. If the stew needs no seasoning, then that's what you have to do, because if you add salt when it doesn't need any, you've ruined it.

I try to maintain what the mixer did. A lot of times they're not really in the ballpark due to their monitoring, so I EQ for clarity more than anything.

When you first run something down, can you hear the final product in your head?
Oh yes, virtually instantly, because for the most part I'm working with music that I know what it's supposed to sound like. Once in a while I'll get an album that's so strange to me because of either the music or what the engineer did that I have no idea what it's supposed to sound like, and I often will pass on it. I'll say, "I just don't hear this. Maybe you should go somewhere where they're glued into what you're doing."

For the most part, I'm fortunate to usually work on things that sound pretty good. I work on most of the recordings from great engineers like Bill Schnee, George Massenburg, Ed Cherney, and Al Schmitt.

These are clients that I'm the one they go to if they have a say in where it's mastered.

Every room has its claim to fame, and mine is that I work on more albums nominated for engineering Grammys than any other room, and probably by a factor of three or four to the next closest room.

How has mastering changed over the years from the time you started until the way it is now?

My answer is maybe different than everyone else's. It hasn't changed at all! In other words, what you're doing is finessing what an engineer and artist has created into its best possible form.

If an engineer says, "I don't know what it is, but the vocal always seems to be a little cloudy," I can go in there and keep his mix the same, yet still make the vocal clearer. That's what I did in 1968, and that's what I still do. The process is the same, and the goal is the same.

I don't master differently for different formats, because you essentially make it sound as proper as you can, and then you transfer it to the final medium using the best equipment.

Do you think that working on vinyl would help a newer mastering engineer who's never had that experience?

I don't know if working on vinyl helps. I think having worked on many different types of music over the years helps.

In one sense, being from the vinyl days I was used to doing all the moves in real time. I always cut directly from the master tapes, so if you blew a fade on the fourth cut, you started over again. So the concept of being able to do everything in real time instead of going into a computer probably affects the way I master now. I don't look at things as, "Oh, I can put this in and fine-tune this and move this up and down." I look at it as to what I can do in real time.

I find that the idea that you have a track for every instrument and you put them all together to have great clarity doesn't work. I think it works the opposite way. The more you separate it, the harder it is to put together and have clarity, so if you're EQing for musical clarity to hear what is down there, that's unchanged today from way back 40 years ago.

It's the same process, and the EQ that would make somebody call up and say, "Wow, I really like it. I can hear everything and yet it's still full," is still as valid today as it was then.

Many artists are willing to spend money on a real mastering engineer to get the ears of a pro. Have you experienced that?

Yes, but I think the caveat is how much money they're willing to spend. The amount of money it takes to open up a mastering facility today is minuscule compared to what it used to be. You can do almost everything without a big investment.

The question then becomes, "Are you willing to spend the extra money for the expertise of the mastering engineer?" Just owning a Pro Tools system does not make you a mastering engineer.

The fact that you have a finely tuned room and a super high-quality playback system is hard to compete with.
Yes, but if you figure in the jobs that no one attends, they can't experience that, so we have to supply something that they can hear at home that's better than what they could do themselves. There are some engineers that do come to our facility, but the majority is being sent to us now.

What's the hardest thing you have to do?
I come from a time when an album had a concept to it. The producer worked with one engineer and one studio, the group recorded everything, and there was cohesiveness as to what was put before you. Once you got what they were doing, you sort of had the album done.

The multiple-producer album to me is the biggest challenge, because you might have three mixes from Nashville, a couple from New York, and two that are really dark and muddy, and three are bright and thin. The only good part that I see about this is that you absolutely have to use a mastering engineer in this case, or the mixes don't work together.

The hard part for the mastering engineer is to find some middle ground, so that the guy with the bright, thin sound is still happy with what he's done and doesn't drive off the road when the dull, thick one plays after the bright, thin one. That's the biggest challenge in mastering—making what is really a cafeteria sound feel like a planned meal.

I'm very proud of the fact that I've trained a lot of good mastering engineers, and I'll tell them, "You're not going to learn how to master working on a Massenburg mix. It's pretty well done, and if he didn't like it, he wouldn't have sent it. When you get mixes from engineers that are not great, or you get these multiple-engineer things, then you can sort of learn the art of mastering by making these things work using your ears."

You started with digital technology from when it first came out. How has it changed?
I get a lot of 96/24 stuff in now, and a 96/24 recording done with great converters can sound terrific.

The truth of the matter is that the tools are getting so much better. Digital technology is moving so fast, and it's gone from, in my view, absolute garbage to "Hey, this is pretty good."

Mastering engineers don't like that because they used to be the only ones that could make a track loud, but the reality is that everyone today can make their mix loud. Once that becomes absolutely no trick at all, then the question becomes, "Are there things that maybe we should do besides just make it loud?"

I'm hoping that there's still going to be a business for someone that treats the music with love and respect when they're mastering it, and I think there's going to be a small reversion away from, "I want the loudest master."

It's fantastic that what you do has weathered the test of time.
Yes. It's still the same concept. I don't master any differently today than I did in 1968. The speakers allow me to put the right stuff on, and if they steer me wrong, then they're worthless.

GLOSSARY

0dB FS (Full Scale). The highest level that can be recorded in the digital domain. Recording beyond 0dBFS can result in distortion.

5.1. A speaker system that uses three speakers across the front and two stereo speakers in the rear, along with a subwoofer.

1630. A first-generation two-track digital tape machine made by Sony utilizing a separate digital processor and a 3/4-inch U-matic video tape machine for storage. The 1630 was the primary master tape delivered to the pressing plant in the early years of the CD, but they are considered obsolete today. A model 1610 predated this machine.

AAC. Advanced Audio Coding is a standard lossy data compression encoding scheme for digital audio used exclusively on Apple Music.

acetate. An acetate is a single-sided vinyl check disc, sometimes called a *ref*. Due to the extreme softness of the vinyl, an acetate has a limited number of plays (five or six) before it wears out. See *ref*.

A/D. Analog-to-digital converter. This device converts the analog waveform into the digital language that can be used by a digital audio workstation.

AIFF. Audio Interchange File Format (also known as *Apple Interchange File Format*) is an audio file format designed for use in the Apple Macintosh operating system but now widely used in PCs as well.

airplay. When a song gets played on the radio.

arrangement. The way the instruments are combined in a song.

asset. A multimedia element, either sound, picture, graphic, or text.

Atmos (see Dolby Atmos)

attack. The first part of a sound envelope. On a compressor/limiter, a control that affects how that device will respond to the attack of a sound.

attenuation. A decrease in gain or level.

Augspurger. George Augspurger of Perception Inc. in Los Angeles is one of the most revered studio designers. He also designs large studio monitors, each having dual 15-inch woofers and a horn tweeter.

automation. A system that memorizes and then plays back the position of all faders and mutes on a console. In a DAW, the automation can also record and play back other parameters, including sends, returns, panning, and plug-in parameters.

bandwidth. The number of frequencies that a device will pass before the signal degrades. A human being can supposedly hear from 20Hz to 20kHz, so the bandwidth of the human ear is 20 to 20kHz. Sometimes applies to computer data rate, where a high rate per second represents a wider bandwidth.

barcode. A series of vertical bars of varying widths in which the numbers 0 through 9 are represented by a different pattern of bars that can be read by a laser scanner. Barcodes are commonly found on consumer products and are used for inventory control, and in the case of CDs, to tally sales.

bass management. A circuit that utilizes the subwoofer in an immersive audio playback system to provide bass extension for the five main speakers. The bass manager steers all frequencies below approximately 100 Hz into the subwoofer along with the LFE source signal. *See LFE.*

bass redirection. Another term for bass management.

big ears. The ability to be very aware of everything going on within the session and with the music.

bit rate. The transmission rate of a digital system.

Blu-ray. The name of the optical disc format initially developed by Sony and Philips (inventor of the compact disc, cassette, and laserdisc) as a next-generation data and video storage format alternative to DVD.

bottom. Bass frequencies, the lower end of the audio spectrum. See also *low end*.

bottom end. *See* bottom.

brick-wall filter. A low-pass filter used in digital audio set to half the sampling rate so the frequency response does not go beyond what is suggested by the Nyquist Theorem.

brick-wall limiter. A limiter employing look-ahead technology that is so efficient that the signal will never exceed a certain predetermined level, and there will be no digital overs.

buss. A signal pathway.

CALM Act. Legislation passed by Congress to ensure that television commercials and programs are broadcast equally loud.

catalog. Older albums or recordings under control of the record label.

clean. A signal with no or barely noticeable distortion.

clip. To overload and cause distortion.

clipping. When an audio signal begins to distort because a circuit in the signal path is overloaded, the top of the waveform becomes "clipped" off and begins to look square instead of rounded. This usually results in some type of distortion, which can vary from soft and barely noticeable to horribly crunchy-sounding.

codec. Code-decode. A codec is a software algorithm that encodes and decodes a particular file format such as FLAC, AAC or MP3.

color. To affect the timbral qualities of a sound.

comb filter. A distortion produced by combining an electronic or acoustic signal with a delayed copy of itself. The result is peaks and dips introduced into the frequency response.

competitive level. A mix level that is as loud as your competitor's mix.

compression. Signal processing that controls the dynamics of a sound.

compressor. A signal-processing device used to control audio dynamics.

cross-fade. In mastering, the overlap of the end of one song into the beginning of the next.

cut. To decrease, attenuate, or make less.

cutter head. The assembly on a vinyl-cutting lathe that holds the cutting stylus between a set of drive coils powered by very high-powered (typically, 1000 to 3500 watts) amplifiers.

D/A. Digital-to-analog converter, sometimes called a DAC. This device converts the digital audio data stream into analog audio.

DAC. Digital-to-analog convertor. The device that converts the signal from the digital domain to the analog domain.

data compression. A process that uses a specially designed algorithm to decrease the number of bits in a file for more efficient storage and transmission.

DAW. A digital audio workstation. A system designed for recording, editing and playback of audio. Modern DAW systems are primarily run using software on computers.

dB. Stands for *decibel*, which is a unit of measurement of sound level or loudness. The smallest change in sound level that an average human can hear is 1dB.

decay. The time it takes for a signal to fall below audibility.

decoupling. Isolating speakers from a desk or console by using rubber or carpet.

DDP. Disc Description Protocol. A proprietary format that is low in errors and allows high-speed glass-master cutting. It is currently the standard delivery format for CDs and DVDs.

digital domain. When a signal source is converted into a series of electronic pulses represented by 1s and 0s, the signal is then in the digital domain.

digital overs. The point beyond 0 on a digital processor where the red over indicator lights, resulting in a digital overload.

'dither. A low-level noise signal used to limit quantization distortion when lowering the bit resolution of an audio file.

Dolby Atmos. A surround sound technology developed by Dolby Laboratories. It expands on existing surround sound systems by adding height channels, allowing sounds to be interpreted as three-dimensional objects.

DSP. Digital Signal Processing. Processing within the digital domain, usually by dedicated microprocessors.

dynamic range. A ratio that describes the difference between the highest and lowest signal level. The higher the number, equaling the greater dynamic range, the better.

edgy. A sound with an abundance of midrange frequencies.

element. A component or ingredient of the mix.

EQ. Equalizer, or to adjust the equalizer (tone controls) to affect the timbral balance of a sound.

equalization. Adjustment of the frequency spectrum to even out or alter tonal imbalances.

equalizer. A tone control that can vary in sophistication from very simple to very complex. See *parametric equalizer*.

feather. Rather than applying a large amount of equalization at a single frequency, small amounts are added at the frequencies adjoining the one of principle concern.

FLAC. Free Lossless Audio Codec. A lossless file format used to make digital audio files smaller in size, yet that suffer no degradation of audio quality.

Fletcher-Munson curves. A set of measurements that describes how the frequency response of the ear changes at different sound-pressure levels. For instance, we generally hear very high and very low frequencies much better as the overall sound pressure level is increased.

flip the phase. Selecting the phase switch on a console, preamp, or DAW channel in order to find the setting with the greatest bass response.

gain. The amount a sound is boosted.

gain reduction. The amount of signal level attenuation as a result of compression or limiting.

gain staging. Adjusting the gain of each processing stage of the signal chain so the output of one doesn't overload the input of another.

glass master. The first and most important step in the CD replication process involving electroplating a glass block in order to make the CD stampers.

groove. The pulse of the song and how the instruments dynamically breathe with it. Or, the part of a vinyl record that contains the mechanical information that is transferred to electronic info by the stylus.

headroom. The amount of dynamic range between the normal operating level and the maximum output level, which is usually the onset of clipping.

Hertz. A measurement unit of audio frequency, defined by the number of cycles per second. High numbers represent high-pitched sounds, and low numbers represent low-pitched sounds.

high end. The high frequency response of a device.

high-pass filter. An electronic circuit that allows the high frequencies to pass while attenuating the low frequencies. Used to eliminate low-frequency artifacts, such as hum and rumble. The frequency point where it cuts off is usually either switchable or variable.

hypercompression. Too much buss compression during mixing or mastering in an effort to make the recording louder results in what's known as hypercompression, a condition that essentially leaves no dynamics and makes the track sound lifeless.

Hz. Short for Hertz. See *Hertz*.

I/O. The input/output of a device.

immersive audio. Multi-dimensional sound that completely envelops the listener because of speakers placed around the listening environment as well as overhead.

ISRC code. The International Standard Recording Code is used to uniquely identifying sound recordings and music video recordings. An ISRC code identifies a particular recording, not the song itself; therefore, different recordings, edits, and remixes of the same song will each have their own ISRC codes.

kbs. Kilobits per second. The amount of digital information sent per second. Sometimes referred to as *bandwidth*.

kHz. One-thousand Hertz (example: 4kHz = 4,000Hz).

knee. How quickly a compressor will turn on once it reaches the threshold. A *soft knee* turns on gradually and is less audible than a *hard knee*.

lacquer. The vinyl master, which is a single-sided 14-inch disc made of aluminum substrate covered with a soft cellulose nitrate. A separate lacquer is required for each side of a record. Since the lacquer can never be played, a ref or acetate is made to check the disc. See *ref* and *acetate*.

latency. Latency is a measure of the time it takes (in milliseconds) for your audio signal to pass through your system during the recording process. This delay is caused by the time it takes for your computer to receive, understand, process, and send the signal back to your outputs.

LFE. Low-frequency effects channel. This is a special channel of 30Hz to 120Hz information primarily intended for special effects, such as explosions in movies. The LFE has an additional 10dB of headroom to accommodate the required sound pressure level of the low frequencies.

limiter. A signal-processing device used to constrict or reduce audio dynamics, reducing the loudest peaks in volume.

LKFS. Stands for Loudness, K-weighted, relative to Full Scale, which distinguishes itself from the normal dBFS peak meters found on all digital gear in that it measures the loudness not instant by instant, but over a period of time.

look-ahead. In a mastering limiter, look-ahead processing delays the audio signal a small amount (about 2 milliseconds or so) so that the limiter can anticipate the peaks in such a way that it catches the peak before it gets by.

lossless compression. A compression format that recovers all the original data from the compressed version and suffers no degradation of audio quality as a result. FLAC and ALAC are lossless compression schemes.

lossy compression. A digital file-compression format that cannot recover all of its original data from the compressed version. Supposedly some of what is normally recorded before compression is imperceptible, with the louder sounds masking the softer ones. As a result, some data can be eliminated since it's not heard anyway. This selective approach, determined by extensive psychoacoustic research, is the basis for lossy compression. MP3 and AAC are lossy compression schemes.

low end. The lower end of the audio spectrum, or bass frequencies usually below 200Hz.

low-pass filter. An electronic frequency filter that allows only the low frequencies to pass while attenuating the high frequencies. The frequency point where it cuts off is usually either switchable or variable.

LPCM. Linear Pulse Code Modulation. This is the most common method of digital encoding of audio used today and is the same digital encoding method used by current audio CDs. In LPCM, the analog waveform is measured at discrete points in time and converted into a digital representation.

LUFS. Stands for Loudness Units relative to Full Scale. This was formerly the European standard for loudness, but is now identical to LKFS.

makeup gain. A control on a compressor/limiter that applies additional gain to the signal. This is required since the signal is automatically decreased when the compressor is working. Makeup gain "makes up" for the lost gain and brings it back to where it was prior to being compressed.

mastering. The process of turning a collection of songs into an album by making them sound like they belong together in tone, volume, and timing (spacing between songs).

metadata. Data that describes the primary data. For instance, metadata can be data about an audio file that indicates the date recorded, sample rate, resolution, artist, record label, publisher, and so on.

midrange. Middle frequencies starting from around 250Hz up to 4,000Hz.

mid-side. A technique for representing stereo information by splitting a stereo signal into two parts: the middle (M) and the left and right sides (S). The mid channel is the sum of the left and right channels, while the side channel is the difference between the two.

modeling. Developing a software algorithm that is an electronic representation of the sound of hardware audio device down to the smallest behaviors and nuances.

monaural. A mix that contains a single channel and usually comes from only one speaker.

mono. Short for monaural, or single audio playback channel.

mother. In either vinyl or CD manufacturing, the intermediate step from which a stamper is made.

MP3. A data-compression format used to make audio files smaller in size.

multi-band compression. A compressor that is able to individually compress different frequency bands as a means of having more control over the compression process.

mute. An on/off switch. To mute something would mean to turn it off.

native resolution. The sample rate and bit depth of a distribution container. For example, the native resolution of a CD is 44.1kHz and 16 bits. The native resolution in film work is 48kHz and 24 bits.

noise shaping. Dither that moves much of the injected noise to a part of the audio spectrum beyond where the ear is less likely to hear it.

normalization. A selection on a DAW that looks for the highest peak of an audio file and adjusts all the levels of the file upward to match that level.

Nyquist Sampling Theorem. A basic tenet of digital audio that states that the frequency response of a system cannot go beyond half the sampling rate. If that occurs, artifacts known as aliasing are introduced into the signal, destroying the purity of the audio.

out of phase. The polarity of two channels (it could be the left and right channels of a stereo program) are reversed, thereby causing the center of the program (like the vocal) to diminish in level. Electronically, when one cable is wired backwards from all the others.

overs. Digital overs occur when the level is so high that it tries to go beyond 0dBFS on a typical digital level meter found in just about all equipment. A red overload indicator usually will turn on, possibly accompanied by audible clipping.

overcut. A record that has grooves that are too wide and deep, which can cause higher noise levels and more crackles and pops

pan. Short for *panorama*, pan indicates the left and right position of an instrument within the stereo spectrum.

panning. Moving a sound across the stereo spectrum.

parametric equalizer. A tone control where the gain, frequency, and bandwidth are all variable.

parts. The different masters sent to the pressing plant. A mastering house may make different parts/masters for CD, cassette, and vinyl, or send additional parts to pressing plants around the world.

peaks. A sound that's temporarily much higher than the sound surrounding it.

phantom image. In a stereo system, if the signal is of equal strength in the left and right channels, the resultant sound appears to come from in between them. This is a phantom image.

phase. The relationship between two separate sound signals when combined into one.

phase meter. A dedicated meter that displays the relative phase of a stereo signal.

phase shift. The process during which some frequencies (usually those below 100Hz) are slowed down ever so slightly as they pass through a device. This is usually exaggerated by excessive use of equalization and is highly undesirable.

pitch. On a record, the velocity of the cutter head. Measured in the number of lines (grooves) per inch.

plug-in. An add-on to a computer application that adds functionality to it. EQ, modulation, and reverb are examples of DAW plug-ins.

PMCD. Pre-Mastered CD, an obsolete format similar to a CD-R except that it has PQ codes written on the lead-out of the disc to expedite replication.

PQ codes. Subcodes included along with the audio data channel as a means of placing control data, such as start IDs and the table of contents, on a CD.

pre-delay. The time between the dry sound and the onset of reverberation. The correct setting of the pre-delay parameter can make a difference in the clarity of the mix.

presence. Accentuated upper-midrange frequencies (anywhere from 4k to 6kHz).

producer. The equivalent of a movie director, the producer has the ability to craft the songs of an artist or band technically, sonically, and musically.

proximity effect. The inherent low-frequency boost that occurs with a directional microphone as it gets closer to the signal source.

Pultec. An tube-based equalizer sold during the '50s and '60s by Western Electric that is highly prized today for its smooth, unique sound.

pumping. When the level of a mix increases and then decreases noticeably. Pumping is caused by the improper setting of the attack and release times on a compressor.

punchy. A description for a quality of sound that infers good reproduction of dynamics with a strong impact. The term sometimes means emphasis in the 200Hz and 5kHz areas.

Q. The bandwidth, or the frequency range of a filter or equalizer.

range. On a gate or expander, a control that adjusts the amount of attenuation that will occur to the signal when the gate is closed.

ratio. A parameter control on a compressor/limiter that determines how much gain reduction will occur when the signal exceeds the threshold.

record. A generic term for the distribution method of a recording. Regardless of whether it's a CD, vinyl, or a digital file, it is still known as a record.

ref. Short for *reference record*, a ref is a single-sided vinyl check disc, sometimes called an *acetate*. Due to the extreme softness of the vinyl, a ref has a limited number of plays (five or six) before it wears out. See *acetate*.

reference level. This is the audio level, either electronic and acoustic, to which a sound system is aligned.

release. The last part of a sound envelope. On a compressor/limiter, a control that affects how that device will respond to the release part of the sound envelope.

resonance. See *resonant frequency*.

resonant frequency. A particular frequency or band of frequencies that is accentuated, usually due to some sympathetic acoustic, electronic, or mechanical factor.

return. Inputs on a recording console especially dedicated for effects devices such as reverbs and delays. The return inputs are usually not as sophisticated as normal channel inputs on a console.

RIAA. Recording Industry Association of America. A trade organization for record labels but dominated by the major labels.

RIAA Curve. An equalization curve instituted by the Recording Industry Association of America (the RIAA) in 1953 that enabled the grooves to be narrowed, thereby allowing more of them to be cut on the record, which increased the playing time and decreased the noise. This was accomplished by boosting the high frequencies by about 17dB at 15kHz and cutting the lows by 17dB at 50Hz when the record was cut. The opposite curve is then applied during playback.

RMS meter. A meter that reads the average level of a signal.

roll-off. Usually another word for high-pass filter, although it can refer to a low-pass filter as well.

sample rate. The rate at which the analog waveform is measured. The more samples per second of the analog waveform that are taken, the better the digital representation of the waveform that occurs, resulting in greater bandwidth for the signal.

sequencing. Setting the order in which the songs will play on a CD or vinyl record.

scope. Short for oscilloscope, an electronic measurement device that produces a picture of the audio waveform.

shelving curve. A type of equalizer circuit used to boost or cut a signal above or below a specified frequency. Usually the high- and low-band equalizers built into many mixing boards are the shelving type.

sibilance. A short burst of high frequencies in a vocal sometimes due to heavy compression, resulting in the S sounds being overemphasized.

solid state. A device that uses solid semiconductors rather that vacuum tubes.

soundfield. The listening area containing mostly direct sound from the monitor speakers.

source. An original master that is not a copy or a clone.

spectrum. The complete audible range of audio signals.

SPL. Sound-pressure level.

spread. The time in between songs on a CD, cassette, or vinyl record.

SRC. Sample-rate conversion.

stamper. In either vinyl or CD manufacturing, a negative copy bolted into the presser to actually stamp out records or CDs.

stem. An instrument group of tracks that make up a full mix. Stems are typically divided into drum stems, bass stem, vocal stem and instruments stem, although they may get further defined, such as background vocal stem, keyboards stem, etc. Each stem contains all of the processing and effects added during the mix.

sub. Short for subwoofer.

subwoofer. A low-frequency speaker with a frequency response from about 25Hz to 120Hz.

tempo. The rate of speed, usually represented in beats per minute, that a song is played.

test tones. A set of tones used to calibrate a playback system. In the days of tape, they were added to a tape to help calibrate the playback machine.

threshold. The point at which an effect takes place. On a compressor/limiter, for instance, the threshold control adjusts the point at which compression will begin.

timbre. Tonal color.

track. A term sometimes used to mean a song. In recording, a separate musical performance that is recorded.

transformer. An electronic component that either matches or changes the impedance. Transformers are large, heavy, and expensive, but are in part responsible for the desirable sound in vintage audio gear.

trim. A control that sets the gain of a device, or the process of reducing the size or playing time of an audio file.

tube. Short for *vacuum tube*; an electronic component used as the primary amplification device in most vintage audio gear. Equipment utilizing vacuum tubes runs hot, is heavy, and has a short life, but it has a desirable sound.

TV mix. A mix without the vocals so the artist can sing live to the backing tracks during a television appearance.

U-matic. An industrial digital-video machine utilizing a cassette storing 3/4-inch tape. The U-matic is the primary storage device for the 1630 digital processor.

underscore. The instrumental score of a movie that plays below the dialogue and/or sound effects.

unity gain. When the output level of a process or processor exactly matches its input level.

variable pitch. On a record, varying the number of grooves per inch depending upon the program material.

Vinylite. The vinyl used to make records actually comes in a granulated form called *Vinylite*. Before being pressed, it is heated into the form of modeling clay and colored with pigment.

WAV. A WAV file is an audio data file developed by the IBM and Microsoft corporations, and is the PC equivalent of an AIFF file. It is identified by the ".wav" file extension.

word length. The number of bits in a word. Word length is in groups of eight. The longer the word length, the better the dynamic range.

ABOUT BOBBY OWSINSKI

Producer/engineer Bobby Owsinski is one of the best selling authors in the music industry with 25 books that are now staples in audio recording, music, and music business programs in colleges around the world, These include The Mixing Engineer's Handbook, The Recording Engineer's Handbook, The Musician's AI Handbook, and more. He's also a contributor to Forbes writing on the new music business, his popular blogs and Inner Circle podcast have won numerous awards, and he's appeared on CNN and ABC News as a music branding and audio expert.

Visit Bobby's music production blog at bobbyowsinskiblog.com, his Music 3.0 music industry blog at music3point0.com, his podcast at bobbyoinnercircle.com, his online courses at bobbyowsinskicourses.com, and his website at bobbyowsinski.com.

BOBBY OWSINSKI BIBLIOGRAPHY

The Mixing Engineer's Handbook 5th Edition (BOMG Publishing)

The Music Mixing Workbook (BOMG Publishing)

The Recording Engineer's Handbook 5th Edition (BOMG Publishing)

The Musican's AI Handbook (BOMG Publishing)

Social Media Promotion For Musicians 3rd Edition - *The Manual For Marketing Yourself, Your Band or your Music Online* (BOMG Publishing)

The Drum Recording Handbook 2nd Edition [with Dennis Moody] (Hal Leonard Publishing)

How To Make Your Band Sound Great (Hal Leonard Publishing)

The Studio Musician's Handbook [with Paul ILL] (Hal Leonard Publishing)

Music 4.1 - A Survival Guide To Making Music In The Internet Age 4th Edition (Hal Leonard Publishing)

The Music Producer's Handbook 2nd Edition (Hal Leonard Publishing)

The Musician's Video Handbook (Hal Leonard Publishing)

Mixing And Mastering With T-Racks: The Official Guide (Course Technology PTR)

The Touring Musician's Handbook (Hal Leonard Publishing)

The Ultimate Guitar Tone Handbook [with Rich Tozzoli] (Alfred Music Publishing)

The Studio Builder's Handbook [with Dennis Moody] (Alfred Music Publishing)

Abbey Road To Ziggy Stardust [with Ken Scott] (Alfred Music Publishing)

The Audio Mixing Bootcamp (Alfred Music Publishing)

Audio Recording Basic Training (Alfred Music Publishing)

Deconstructed Hits: Classic Rock Vol. 1 (Alfred Music Publishing)

Deconstructed Hits: Modern Pop & Hip-Hop (Alfred Music Publishing)

Deconstructed Hits: Modern Rock & Country (Alfred Music Publishing)

The PreSonus StudioLive Mixer Official Manual (Alfred Music Publishing)

You can get more info and read excerpts from each book by visiting the excerpts section of bobbyowsinski.com.

BOBBY OWSINSKI LINKEDIN LEARNING VIDEO COURSES

The Audio Mixing Bootcamp Video Course

Audio Recording Techniques

Audio Mastering Techniques

Music Studio Setup and Acoustics

BOBBY OWSINSKI ONLINE COURSES

Available at BobbyOwsinskiCourses.com	Top 40 Mixing Secrets
Vocal Mixing Techniques	Music Mixing Accelerator
The Music Mixing Primer	Fully Booked

BOBBY OWSINSKI'S ONLINE CONNECTIONS

Website: bobbyowsinski.com

Courses: bobbyowsinskicourses.com

Podcast: boobbyoinnercircle.com

Music Production Blog: bobbyowsinskiblog.com

Music Industry Blog: music3point0.com

Forbes Blog: forbes.com/sites/bobbyowsinski/

Facebook: facebook.com/bobby.owsinski

YouTube: youtube.com/polymedia

Pinterest: pinterest.com/bobbyowsinski/

Linkedin: linkedin.com/in/bobbyo

X: @bobbyowsinski

Thanks So Much For Purchasing This Book!

Here's An Extra Free Bonus

PDF Download
Download It Here

(use the QR Code or go to bobbyowsinskicourses.com/masteringreg)

An Instructor Resource Kit for this book is available to qualified instructors

Everything you need to add The Mastering Engineer's Handbook to your course right now!

Each kit includes:

- Syllabus
- Topics for Demonstrations and Discussions for each chapter
- Test Bank and answer key for 12 week semester
- Powerpoint and Keynote presentations for each chapter

The Mastering Engineer's Handbook Instructor Resource Kit is **free to qualified instructors using this book in their in music, recording or production courses.**

Send an email to office@bobbyowsinski.com to receive the download link.